OLYMPUS

DIGITAL CAMERA

E-M1 Mark II

Instruction Manual

Model No. : IM002

- Thank you for purchasing an Olympus digital camera. Before you start to use your new camera, please read these instructions carefully to enjoy optimum performance and a longer service life. Keep this manual in a safe place for future reference.
- We recommend that you take test shots to get accustomed to your camera before taking important photographs.
- The screen and camera illustrations shown in this manual were produced during the development stages and may differ from the actual product.
- If there are additions and/or modifications of functions due to firmware update for the camera, the contents will differ. For the latest information, please visit the Olympus website.

■ **This notice concerns the supplied flash unit and is chiefly directed to users in North America.**

Information for Your Safety

IMPORTANT SAFETY INSTRUCTIONS

When using your photographic equipment, basic safety precautions should always be followed, including the following:

- Read and understand all instructions before using.
- Close supervision is necessary when any flash is used by or near children. Do not leave flash unattended while in use.
- Care must be taken as burns can occur from touching hot parts.
- Do not operate if the flash has been dropped or damaged - until it has been examined by qualified service personnel.
- Let flash cool completely before putting away.
- To reduce the risk of electric shock, do not immerse this flash in water or other liquids.
- To reduce the risk of electric shock, do not disassemble this flash, but take it to qualified service personnel when service or repair work is required. Incorrect reassembly can cause electric shock when the flash is used subsequently.
- The use of an accessory attachment not recommended by the manufacturer may cause a risk of fire, electric shock, or injury to persons.

SAVE THESE INSTRUCTIONS

Indications used in this manual

The following symbols are used throughout this manual.

Tips	Useful information and hints that will help you get the most out of your camera.
☞	Reference pages describing details or related information.

Table of Contents

Playback 76

Menu functions 85

Quick task index

Shooting

Shooting in a mode that is good for image editing	▶	🎥 Picture Mode (🎥 Specification Settings)	100
Recording audio using an external recorder	▶	Linking to external devices	104
Zooming in on photos to check focus	▶	Auto ▶ (Rec View)	109
Extending the useful life of the battery	▶	Quick Sleep Mode	122
Checking battery usage and status	▶	Battery Status	122

Playback/Retouch

			☞
Shooting by outputting to the external monitor	▶	HDMI	117
Brightening shadows	▶	Shadow Adj (JPEG Edit)	106
Dealing with red-eye	▶	Redeye Fix (JPEG Edit)	106
Extracting still images from a 4K movie	▶	In-Movie Image Capture	107
Trimming unwanted sections of movies	▶	Movie Trimming	108
Transferring images to a smartphone	▶	Transferring images to a smartphone	136
Adding location data to images	▶	Adding location data to images	137

Others

			☞
Saving the settings	▶	Assign to Custom Mode	87
Changing the menu display language	▶	Changing the display language	109
Turning off the auto focus sound	▶	Beep sound	117

Others - Basic Shooting

			☞
Taking pictures with the best finish/ Taking black and white pictures	▶	Picture Mode	61, 88
		Art Filter (ART)	33

Names of parts

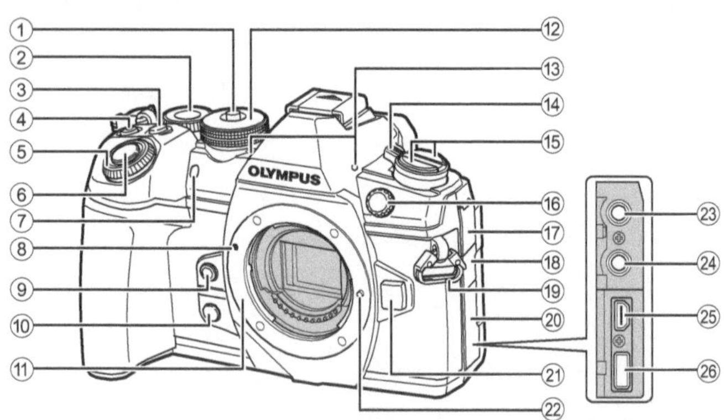

① Mode dial lock................................P. 24
② Rear dial* (⊙)
........................P. 26–29, 42, 70, 78, 128
③ **Fn2** button......................................P. 70
④ ◉ (Movie)/☑ button..............P. 36/P. 82
⑤ Front dial* (⊙)
.........................P. 26–29, 39, 42, 70, 78
⑥ Shutter button.................................P. 25
⑦ Self-timer lamp/AF illuminator
....P. 46, 54/P. 112
⑧ Lens attachment mark....................P. 16
⑨ ◙ (One-touch white balance) button
....P. 67
⑩ ◘ (Preview) button.........................P. 67
⑪ Mount (Remove the body cap before attaching the lens.)
⑫ Mode dial..P. 24
⑬ Stereo microphone ....P. 83, 103, 107
⑭ **ON/OFF** lever................................P. 18
⑮ ⊖ button
AF[•] (AF/Metering mode) button
....P. 43, 45
❏▯ૅ**HDR** (Sequential shooting/ Self-timer/HDR) button ....P. 46, 49, 91
⑯ External flash terminal..................P. 154
⑰ Microphone jack cover
⑱ Headphone jack cover
⑲ Strap eyelet.....................................P. 12
⑳ Connector cover
㉑ Lens release button........................P. 17
㉒ Lens lock pin
㉓ Microphone jack (A commercially available microphone can be connected. ø3.5 stereo mini plug)...................P. 104
㉔ Headphone jack (A commercially available headphones can be connected. ø3.5 stereo pin jack)
㉕ HDMI connector (type D)..............P. 130
㉖ USB connector (type C)
....P. 104, 139, 142

* In this manual, the ⊙ and ⊙ icons represent operations performed using the front dial and rear dial.

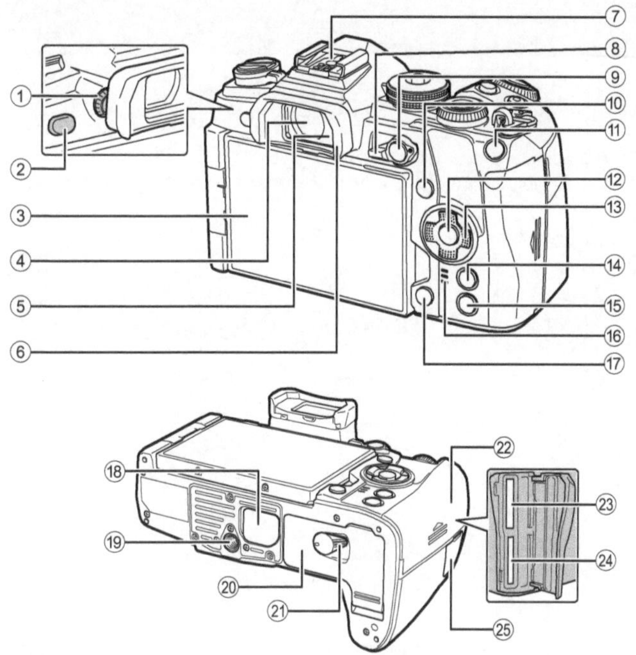

① Diopter adjustment dial ....P. 22
② |◻| (**LV**) button ....P. 22, 68
③ Monitor (Touch screen) ....P. 20, 22, 35, 50, 76, 84
④ Viewfinder ....P. 20, 22, 133
⑤ Eye sensor
⑥ Eyecup ....P. 155
⑦ Hot shoe ....P. 57, 152
⑧ **Fn** lever ....P. 26–29, 124
⑨ **AEL/AFL** button ....P. 45, 81, 123
⑩ **INFO** button ....P. 23, 77
⑪ **Fn1** button.................................P. 40, 78
⑫ Ⓞ button ....P. 50, 78, 85
⑬ Arrow pad* ....P. 78
⑭ **MENU** button ....P. 85
⑮ ▶ (Playback) button.......................P. 78
⑯ Speaker
⑰ 🗑 (Erase) button ....P. 82
⑱ PBH cover......................................P. 150
⑲ Tripod socket
⑳ Battery compartment cover.............P. 13
㉑ Battery compartment lock...............P. 13
㉒ Card compartment cover ....P. 15
㉓ Card slot 1 (UHS-II compatible)......P. 15
㉔ Card slot 2 (UHS-I compatible).......P. 15
㉕ Remote cable terminal cover (Remote cable terminal) ....P. 155

* In this manual, the △▽◁▷ icons represent operations performed using the arrow pad.

1 Preparation

Unpacking the box contents

The following items are included with the camera.
If anything is missing or damaged, contact the dealer from whom you purchased the camera.

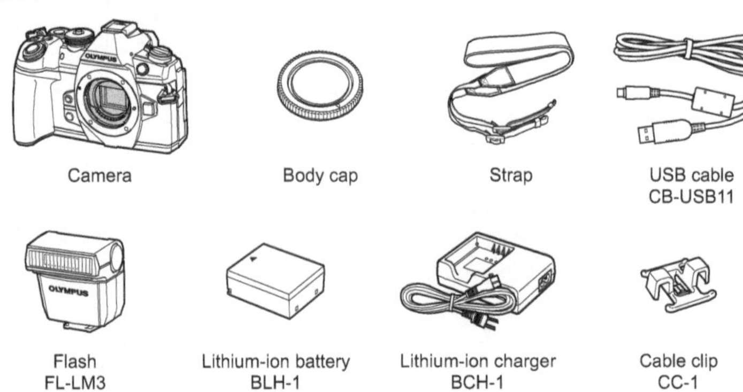

Attaching the strap

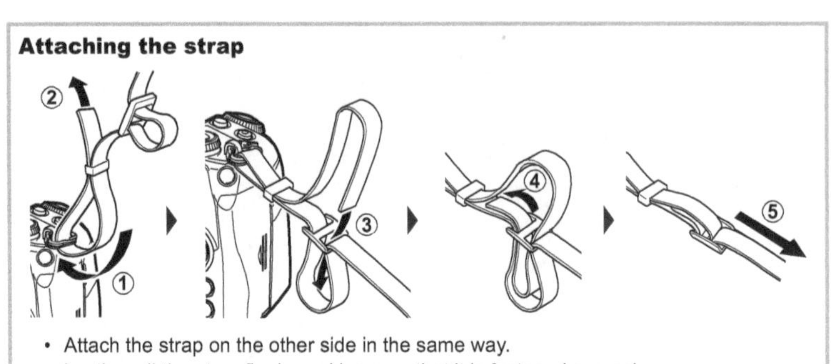

- Attach the strap on the other side in the same way.
- Lastly, pull the strap firmly, making sure that it is fastened securely.

Attaching the cable clip

Use the cable clip to secure the cable, then attach it to the strap.
The cable clip can also be attached to the strap eyelet.

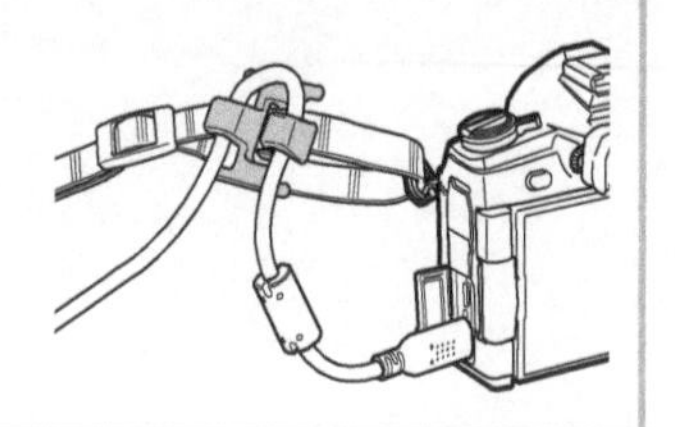

Charging and inserting the battery

1 Charge the battery.

Charging indicator

Charging in progress	Less than 50%	Blinks in orange 1 time/second
	50% or more Less than 80%	Blinks in orange 2 times/second
	80% or more Less than 100%	Blinks in orange 3 times/second
Charging complete		Lights up in green
Charging error		Blinks in green 5 times/second

(Charging time: Approximately 2 hours)

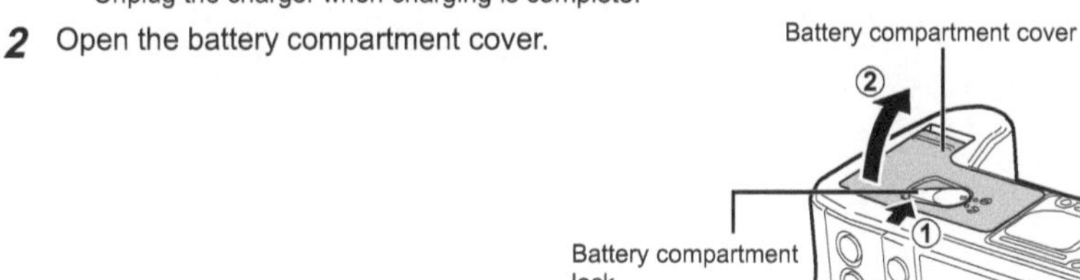

- Unplug the charger when charging is complete.

2 Open the battery compartment cover.

Battery compartment cover
②
Battery compartment lock
①

3 Loading the battery.

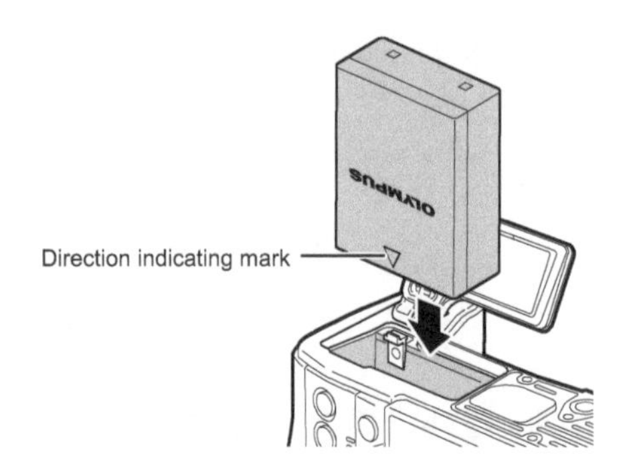

Removing the battery

Turn off the camera before opening or closing the battery compartment cover. To remove the battery, first push the battery lock knob in the direction of the arrow and then remove.

- Contact an authorized distributor or service center if you are unable to remove the battery. Do not use force.

- It is recommended to set aside a backup battery for prolonged shooting in case the battery in use drains.
- Also read "Battery and charger" (P. 146).

Inserting the card

The following types of SD memory card (commercially available) can be used with this camera: SD, SDHC, SDXC, and Eye-Fi.

> **Eye-Fi cards**
> Read "Usable cards" (P. 147) before use.

1 Open the card compartment cover.

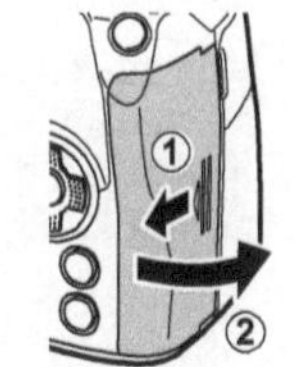

2 Slide the card in until it is locked into place.

- Shooting data will be recorded according to the selected option for [📷 Save Settings] (P. 54).
- ☞ "Usable cards" (P. 147)
- Turn off the camera before loading or removing the card.
- Do not forcibly insert a damaged or deformed card. Doing so may damage the card slot.

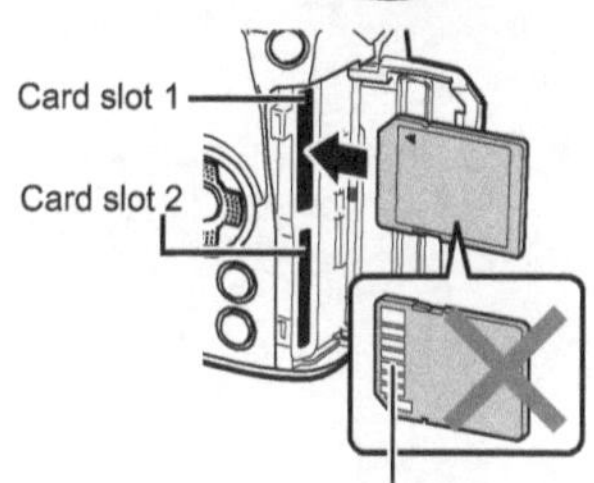

3 Close the card compartment cover.

- Close it securely until you hear it click.
- Be sure the card compartment cover is closed before using the camera.

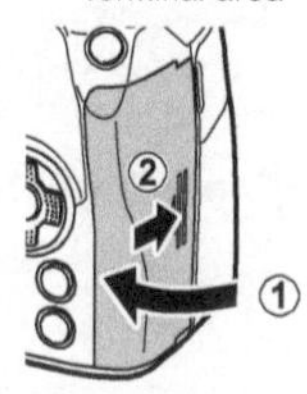

- Cards must be formatted with this camera before first use (P. 110).
- If a UHS-II card is inserted into the slot 2 (UHS-I compatible), the card operates as UHS-I.
- Multiple Eye-Fi cards cannot be used at the same time.

■ Removing the card

Press the card in to eject it. Pull out the card.

- Do not remove the battery or card while the card write indicator (P. 21) is displayed.

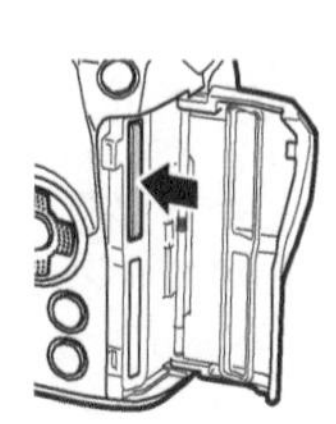

Attaching a lens to the camera

1 Remove the rear cap of the lens and the body cap of the camera.

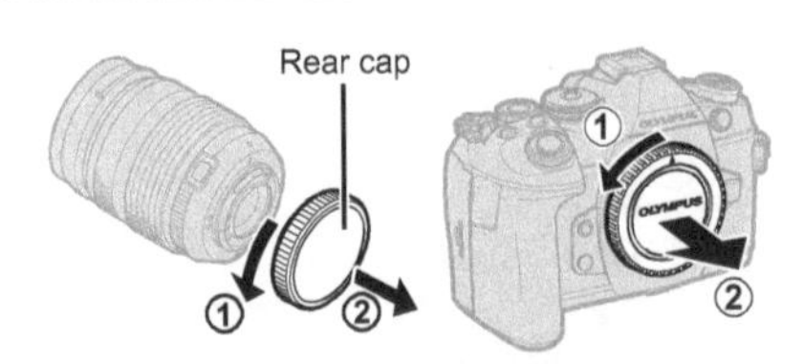

2 Align the lens attachment mark (red) on the camera with the alignment mark (red) on the lens, then insert the lens into the camera's body.

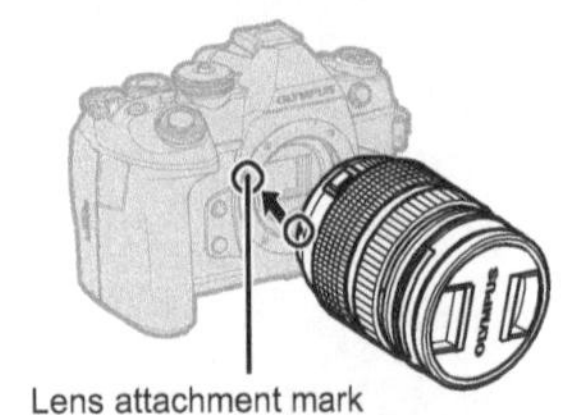

3 Rotate the lens clockwise until you hear it click (direction indicated by arrow ③).

- Make sure the camera is turned off when attaching or removing the lens.
- Do not press the lens release button.
- Do not touch internal portions of the camera.

■ Removing the lens cap

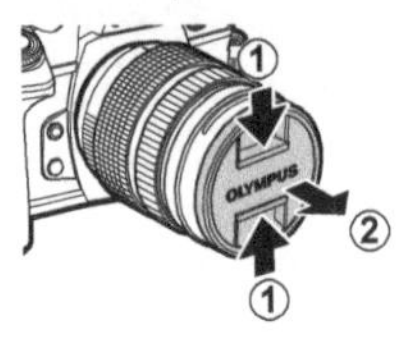

■ Removing the lens from the camera

Make sure the camera is turned off when removing the lens. While pressing the lens release button, rotate the lens in the direction of the arrow.

Interchangeable lenses

Read "Interchangeable lenses" (P. 149).

Using the monitor

You can change the orientation and angle of the monitor.

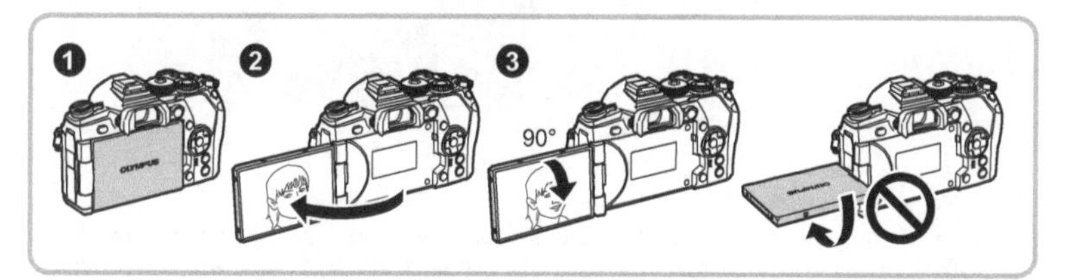

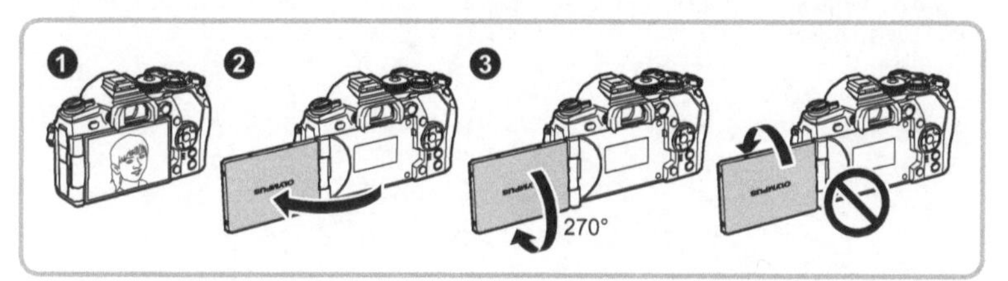

- Rotate the monitor gently within the limits shown. Do not use force; failure to observe this precaution could damage the connectors.
- If you are using a power zoom lens, it will automatically move to the wide angle side when the monitor is set in the selfie position.
- When the monitor is in the selfie position, you can switch to the screen for taking self portraits. ☞ "Shooting self-portraits using the selfie assist menu" (P. 129)

Turning the camera on

1 Set the **ON/OFF** lever to the **ON** position.

- When the camera is turned on, the monitor will turn on.
- To turn the camera off, return the lever to the **OFF** position.

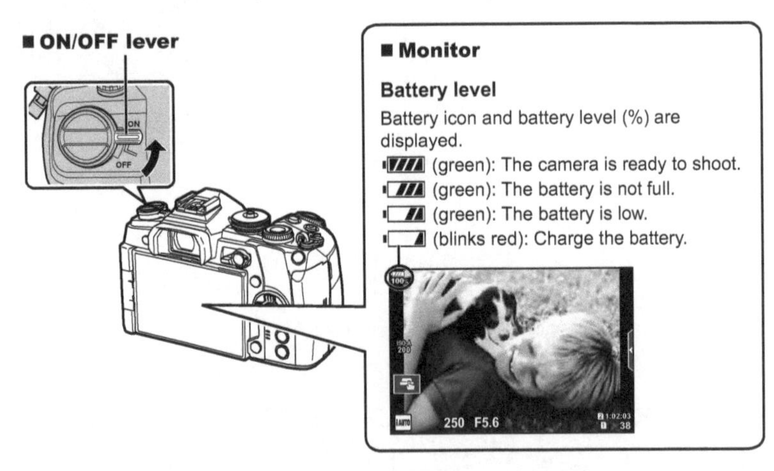

■ Monitor

Battery level

Battery icon and battery level (%) are displayed.

- (green): The camera is ready to shoot.
- (green): The battery is not full.
- (green): The battery is low.
- (blinks red): Charge the battery.

Camera sleep operation

If no operations are performed for a minute, the camera enters "sleep" (stand-by) mode to turn off the monitor and to cancel all actions. The camera activates again when you press any button (the shutter button, [▶] button, etc.). The camera will turn off automatically if left in sleep mode for 4 hours. Turn the camera on again before use.

Setting the date/time

Date and time information is recorded on the card together with the images. The file name is also included with the date and time information. Be sure to set the correct date and time before using the camera. Some functions cannot be used if the date and time have not been set.

1 Display the menus.

- Press the **MENU** button to display the menus.

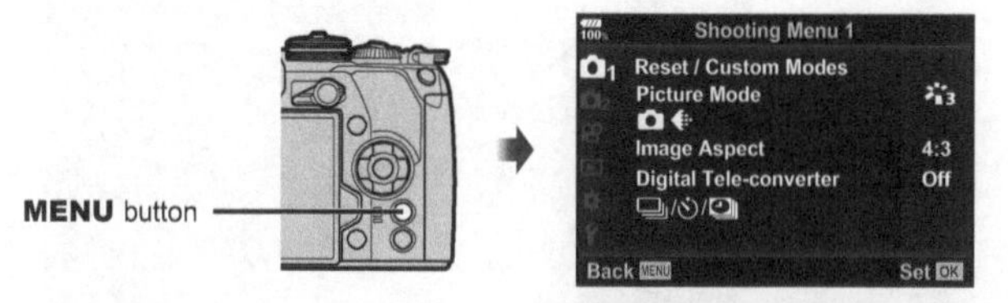

2 Select [◷] in the [🔧] (setup) tab.

- Use △▽ on the arrow pad to select [🔧] and press ▷.
- Select [◷] and press ▷.

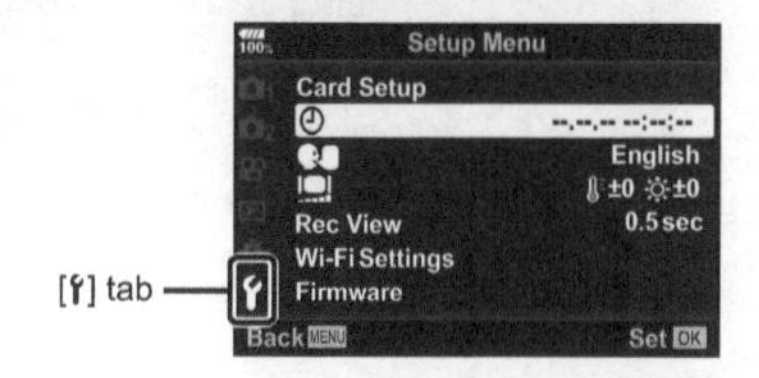

3 Set the date, time, and date format.

- Use ◁▷ to select items.
- Use △▽ to change the selected item.

4 Select [💬] (changing the display language) in the [🔧] (setup) tab.

- You can change the language used for the on-screen display and error messages from English to another language.

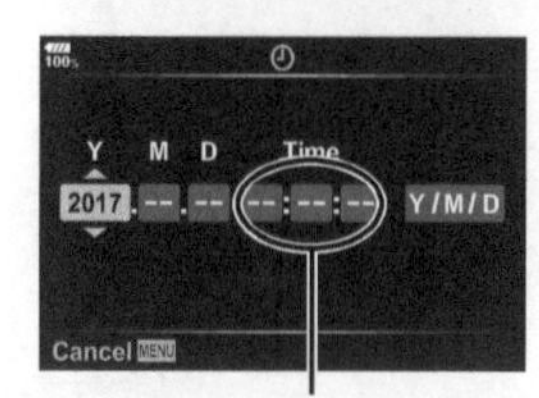

The time is displayed using a 24-hour clock.

5 Save settings and exit.

- Press the ⓞ button to set the camera clock and exit to the menu.
- Press the **MENU** button to exit the menus.

- If the battery is removed from the camera and the camera is left for a while, the date and time may be reset to the factory default setting.

Information displays while shooting

Monitor display during still photography

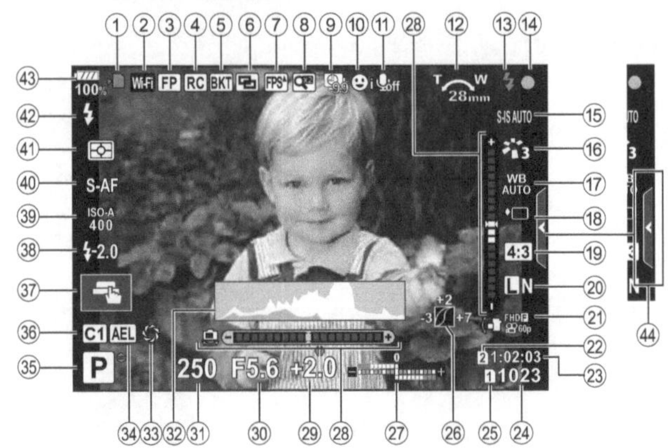

Monitor display during movie mode

① Card write indicator........................P. 15
② Wireless LAN connection.....P. 134–138
③ Super FP flash.............................P. 152
④ RC mode......................................P. 153
⑤ Auto bracket/HDR.................P. 91/P. 49
⑥ Multiple exposure...........................P. 95
Keystone compensation ....P. 97
⑦ High frame rate.............................P. 115
S-OVF*....................................P. 121
⑧ Digital Tele-converter......................P. 88
⑨ Time lapse shooting........................P. 90
⑩ Face priority/Eye priority.................P. 40
⑪ Movie sound ....P. 103
⑫ Zoom operation direction/
Focal length/Internal temperature
warning °C/°F ....P. 164
⑬ Flash ....P. 57
(blinks: charging in progress, lights up: charging completed)
⑭ AF confirmation mark.....................P. 25
⑮ Image stabilizer..............................P. 53
⑯ Picture mode............................P. 61, 88
⑰ White balance..........................P. 42, 52
⑱ Sequential shooting/Self-timer/
Anti-Shock shooting/Silent shooting/
Pro Capture shooting/
High resolution shooting....P. 46–48, 54

* Shown in viewfinder only.

⑲ Aspect ratio....................................P. 54
⑳ Image quality (still images).......P. 55, 88
㉑ Record mode (movies) ....P. 56
㉒ Save slot ....P. 132
㉓ Available recording time ....P. 148
㉔ Number of storable still pictures ...P. 148
㉕ Save Settings ....P. 54
㉖ Highlight & shadow control ....P. 66
㉗ Top: Flash intensity control ....P. 60
Bottom: Exposure compensation....P. 39
㉘ Level gauge ....P. 23
㉙ Exposure compensation value........P. 39
㉚ Aperture value .... P. 26–29
㉛ Shutter speed .... P. 26–29
㉜ Histogram ....P. 23
㉝ Preview...P. 67
㉞ AE lock..................................P. 45, 123
㉟ Shooting mode........................ P. 24–37
㊱ Custom mode ....P. 35, 87
㊲ Touch operation ....P. 35
㊳ Flash intensity control.....................P. 60
㊴ ISO sensitivity...........................P. 42, 51
㊵ AF mode ....P. 43, 51
㊶ Metering mode..........................P. 45, 51
㊷ Flash mode ....P. 57
㊸ Battery level ....P. 18
㊹ Live guide recall.............................P. 31

㊺ Recording level meter...................P. 103
㊻ Silent shooting tab ....P. 38
㊼ Movie (exposure) mode................P. 102
㊽ Movie effect ....P. 37
㊾ Time code ....P. 101

2 Shooting

Switching between displays

The camera is equipped with an eye sensor, which turns the viewfinder on when you put your eye to the viewfinder. When you take your eye away, the sensor turns the viewfinder off and turns the monitor on.

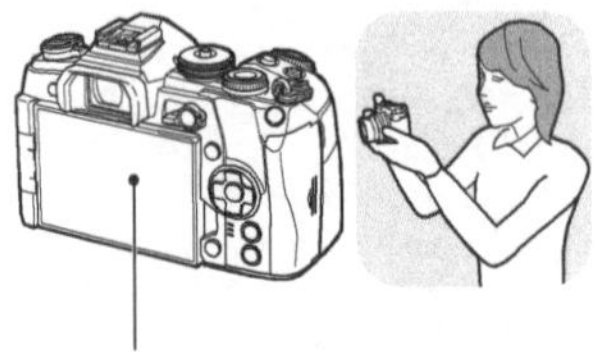

Monitor

Put your eye to the viewfinder

Viewfinder

Live view is displayed on the monitor.

The viewfinder turns on automatically when you bring it to your eye. When the viewfinder is lit up, the monitor turns off.

- The viewfinder will not turn on when the monitor is tilted.
- If the viewfinder is not in focus, put your eye to the viewfinder and focus the display by rotating the diopter adjustment dial.

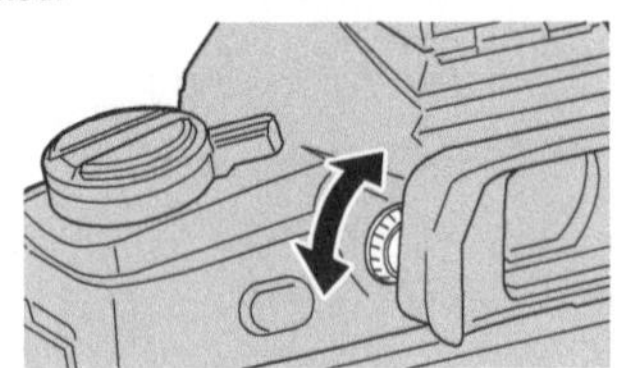

- Press the |O| button to switch between live view and viewfinder shooting (live view and super control panel display). If the super control panel (P. 50) is displayed in the monitor, the viewfinder will turn on when you put your eye to the viewfinder.
- You can display the EVF Auto Switch setting menu if you press and hold the |O| button. ☞ [EVF Auto Switch] (P. 121)

Switching the information display

You can switch the information displayed in the monitor during shooting using the **INFO** button.

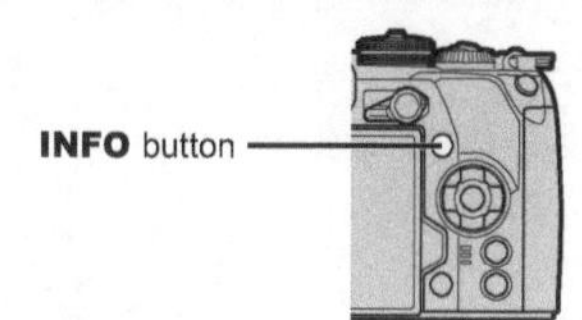

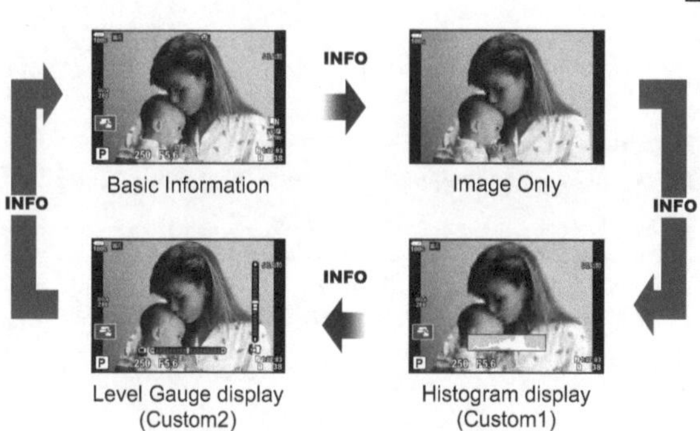

- You can change Custom1 and Custom2 settings. ☞ [▣/Info Settings] > [LV-Info] (P. 115), [▭ Info Settings] (P. 121)
- The information shown in movie (🎥) mode can differ from that displayed in still photography mode. ☞ Video Menu > [🎥 Display Settings] (P. 101)
- The information display screens can be switched in either direction by rotating the dial while pressing the **INFO** button.

Histogram display

A histogram showing the distribution of brightness in the image is displayed. The horizontal axis gives the brightness, the vertical axis the number of pixels of each brightness in the image. Areas above the upper limit at shooting are displayed in red, those below the lower limit in blue, and those within the spot metering range in green.

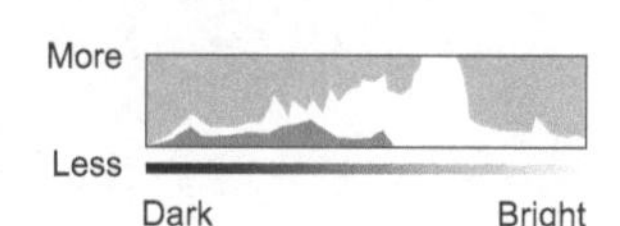

Level gauge display

The orientation of the camera is indicated. The “tilt” direction is indicated on the vertical bar and the “horizon” direction on the horizontal bar.
Use the indicators on the level gauge as a guide.

Shooting still pictures

Use the mode dial to select the shooting mode, and then shoot the image.

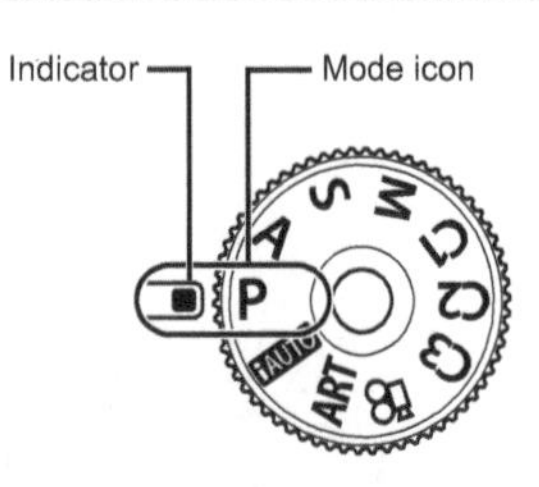

■ **Types of shooting modes**

For how to use the various shooting modes, see the following.

P	P. 26	iAUTO	P. 31
A	P. 27	**ART**	P. 33
S	P. 28	(Movie)	P. 37
M	P. 29	**C1/C2/C3**	P. 35

1 Press the mode dial lock to unlock it, and then turn to set the mode you wish to use.

- When the mode dial lock has been pressed down, the mode dial is locked. Each time you press the mode dial lock, it switches between locked/released.

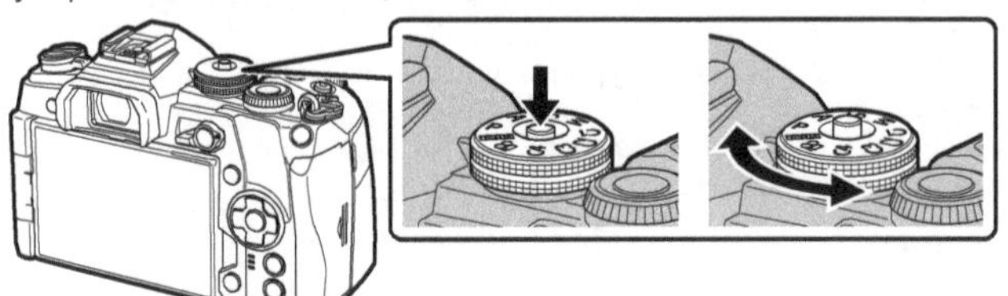

2 Frame the shot.

- Be careful that your fingers or the camera strap do not obstruct the lens or AF illuminator.

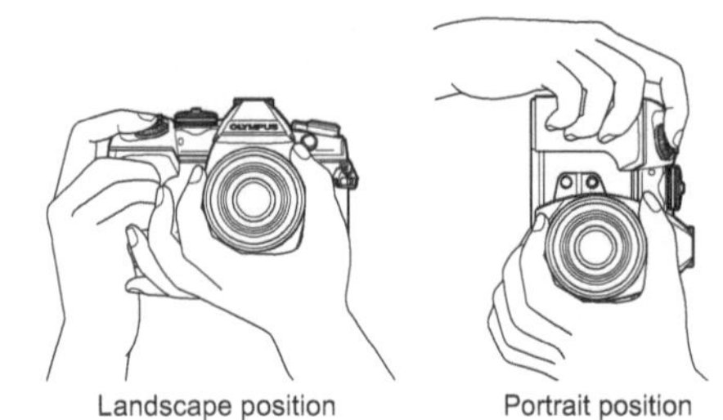

Landscape position Portrait position

3 Adjust the focus.

- Display the subject in the center of the monitor, and lightly press the shutter button down to the first position (press the shutter button halfway).
  The AF confirmation mark (●) will be displayed, and a green frame (AF target) will be displayed in the focus location.

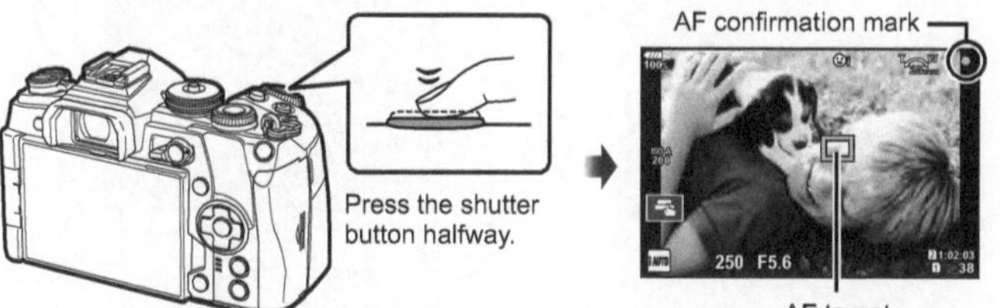

- If the AF confirmation mark blinks, the subject is not in focus. (P. 160)

4 Release the shutter.

- Press the shutter button all the way (fully).
- The camera will release the shutter and take a picture.
- The shot image will be displayed on the monitor.

- You can focus and take pictures using touch controls. ☞ "Shooting with touch screen operations" (P. 35)

Pressing the shutter button halfway and all the way down

The shutter button has two positions. The act of lightly pressing the shutter button to the first position and holding it there is called "pressing the shutter button halfway," that of pressing it all the way down to the second position "pressing the shutter button all (or the rest of) the way down."

Letting the camera choose aperture and shutter speed (Program mode)

P mode is a shooting mode where the camera automatically sets the optimal aperture and shutter speed according to subject brightness. Set the mode dial to **P**.

Aperture value
Shutter speed
Shooting mode

- The shutter speed and aperture selected by the camera are displayed.
- The functions that can be set using the dial differ depending on the **Fn** lever position.

Dial	**Fn lever position**	
	1	2
Front dial	Exposure compensation	ISO
Rear dial	Program shift	White balance

- The shutter speed and aperture value displays will blink if the camera is unable to achieve correct exposure.

Warning display example (blinking)	Status	Action
60" F2.8	The subject is too dark.	• Use the flash.
8000 F22	The subject is too bright.	• The metered range of the camera is exceeded. A commercially available ND filter (for adjusting the amount of light) is required.

- The aperture value at the moment when its indication blinks varies with the lens type and focal length of the lens.
- When using a fixed [ISO] setting, change the setting. ☞ "Changing ISO sensitivity (ISO)" (P. 42, 51)

Program shift (Ps)

In **P** mode, you can choose different combinations of aperture value and shutter speed without altering exposure. "**s**" is displayed next to the shooting mode when the program shift is enabled. To cancel the program shift, rotate the dial until "**s**" is no longer displayed.

Program shift

Choosing aperture (Aperture Priority mode)

A mode is a shooting mode where you choose the aperture and let the camera automatically adjust to the appropriate shutter speed. Set the mode dial to **A** to set the aperture value. Larger apertures (lower F-numbers) decrease depth of field (the area in front of or behind the focus point that appears to be in focus), softening background details. Smaller apertures (higher F-numbers) increase depth of field.

Aperture value

- The functions that can be set using the dial differ depending on the **Fn** lever position.

Dial	Fn lever position	
	1	2
	Exposure compensation	ISO
	Aperture value	White balance

Setting the aperture value

Decreasing aperture value ← → Increasing aperture value

F2.8← F4.0← **F5.6** →F8.0 →F11

- The shutter speed display will blink if the camera is unable to achieve correct exposure.

Warning display example (blinking)	Status	Action
60" F5.6	The subject is underexposed.	• Decrease the aperture value.
8000 F5.6	The subject is overexposed.	• Increase the aperture value. • If the warning display does not disappear, the metered range of the camera is exceeded. A commercially available ND filter (for adjusting the amount of light) is required.

- The aperture value at the moment when its indication blinks varies with the lens type and focal length of the lens.
- When using a fixed [ISO] setting, change the setting. ☞ "Changing ISO sensitivity (ISO)" (P. 42, 51)

Choosing shutter speed (Shutter Priority mode)

S mode is a shooting mode where you choose the shutter speed and let the camera automatically adjust to the appropriate aperture value. Set the mode dial to **S** to set the shutter speed. A fast shutter speed can freeze a fast action scene without any blur. A slow shutter speed will blur a fast action scene. This blurring will give the impression of dynamic motion.

Shutter speed

- The functions that can be set using the dial differ depending on the **Fn** lever position.

Dial	Fn lever position	
	1	2
Front dial	Exposure compensation	ISO
Rear dial	Shutter speed	White balance

Setting the shutter speed

Slower shutter speed ← → Faster shutter speed

60"← 15← 30← **60** →125 →250 →8000

- The aperture value display will blink if the camera is unable to achieve correct exposure.

Warning display example (blinking)	Status	Action
2000 F2.8	The subject is underexposed.	• Set the shutter speed slower.
125 F22	The subject is overexposed.	• Set the shutter speed faster. • If the warning display does not disappear, the metered range of the camera is exceeded. A commercially available ND filter (for adjusting the amount of light) is required.

- The aperture value at the moment when its indication blinks varies with the lens type and focal length of the lens.
- When using a fixed [ISO] setting, change the setting. ☞ "Changing ISO sensitivity (ISO)" (P. 42, 51)

Choosing aperture and shutter speed (Manual mode)

M mode is a shooting mode where you choose both the aperture value and shutter speed. Bulb, time, and live composite photography are also available. Set the mode dial to **M** to set the aperture value and the shutter speed.

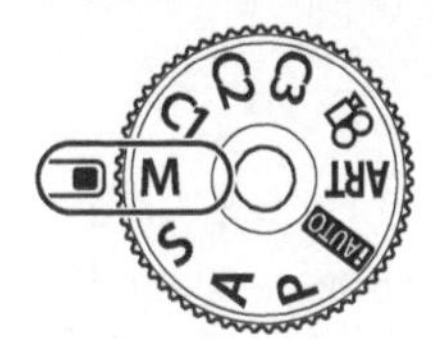

Difference from correct exposure

- The functions that can be set using the dial differ depending on the **Fn** lever position.

Dial	**Fn** lever position	
	1	2
	Aperture value	Exposure compensation*
	Shutter speed	ISO

* When [AUTO] is selected for [ISO], exposure compensation can be adjusted.

Exposure compensation

Difference between exposure setting and exposure with exposure compensation

- The exposure determined by the aperture value and shutter speed you have set and the difference from the suitable exposure measured by the camera are displayed on the monitor.
- Shutter speed can be set to values between 1/8000 and 60 seconds or to [BULB], [LIVE TIME], or [LIVECOMP].
- If you change the aperture value and shutter speed, the brightness of the display on the monitor (or viewfinder) will not change. To display the image as it is going to be shot, set [Live View Boost] (P. 115) in Custom Menu.
- Even when you have set [Noise Reduct.], noise and/or light spots may still be noticeable in the image displayed on the monitor and shot images under certain environmental conditions (temperature etc.) and camera settings.

Noise in images

While shooting at slow shutter speeds, noise may appear on screen. These phenomena occur when the temperature rises in the image pickup device or image pickup device internal drive circuit, causing current to be generated in those sections of the image pickup device that are not normally exposed to light. This can also occur when shooting with a high ISO setting in a high-temperature environment. To reduce this noise, the camera activates the noise reduction function. ☞ [Noise Reduct.] (P. 118)

Shooting with long exposure time (BULB/LIVE TIME)

You can use the BULB/LIVE TIME function for shooting scenes that require long exposure such as night landscapes and fireworks. In **M** mode, set the shutter speed to [BULB] or [LIVE TIME].

Bulb photography (BULB):	The shutter remains open while the shutter button is pressed. The exposure ends when the shutter button is released.
Time photography (LIVE TIME):	The exposure begins when the shutter button is pressed all the way down. To end the exposure, press the shutter button all the way down again.

- During bulb or time photography, the screen brightness changes automatically. ☞ [Bulb/Time Monitor] (P. 118)
- When using [LIVE TIME], the progress of the exposure will be displayed in the monitor during shooting. The display can also be refreshed by pressing the shutter button halfway.
- [Live Bulb] (P. 118) can be used to display the image exposure during bulb photography.
- [BULB] and [LIVE TIME] are not available at some ISO sensitivity settings.
- To reduce camera blur, mount the camera on a sturdy tripod and use a remote cable (P. 155).
- During shooting, there are limits on the settings for the following functions.
  Sequential shooting, self-timer shooting, time lapse shooting, AE bracket shooting, image stabilizer, flash bracketing, multiple exposure*, etc.
  * When an option other than [Off] is selected for [Live Bulb] or [Live Time] (P. 118)
- [Image Stabilizer] (P. 53) turns off automatically.

Live composite photography (dark and light field composite)

You can record a composite image from multiple shots while observing changes in bright flashes of light, such as fireworks and stars, without changing the brightness of the background.

1. Set an exposure time to be the reference in [Composite Settings] (P. 118).
2. In **M** mode, set the shutter speed to [LIVECOMP].
   - When the shutter speed is set to [LIVECOMP], you can display the [Composite Settings] by pressing the **MENU** button.
3. Press the shutter button once to prepare for shooting.
   - You can shoot when a message that indicates preparations are complete is displayed in the monitor.
4. Press the shutter button.
   - Live composite shooting begins. A composite image is displayed after each reference exposure time, allowing you to observe changes in light.
   - During composite shooting, the screen brightness changes automatically. ☞ [Bulb/Time Monitor] (P. 118)

5 Press the shutter button to end shooting.

- The maximum length of composite shooting is 3 hours. However, the available shooting time will vary depending on shooting conditions, and charging state of the camera.

- There are limits on the available ISO sensitivity settings.
- To reduce camera blur, mount the camera on a sturdy tripod and use a remote cable (P. 155).
- During shooting, there are limits on the settings for the following functions.
  Sequential shooting, self-timer shooting, time lapse shooting, AE bracket shooting, image stabilizer, flash bracketing, etc.
- [Image Stabilizer] (P. 53) turns off automatically.

Letting the camera choose the settings (iAUTO mode)

The camera adjusts settings to suit the scene; all you have to do is press the shutter button.
Use live guides to easily adjust such parameters as color, brightness, and background blurring.

1 Set the mode dial to iAUTO.

2 Tap the tab to display the live guides.

- Tap a guide item to display the slider of the level bar.

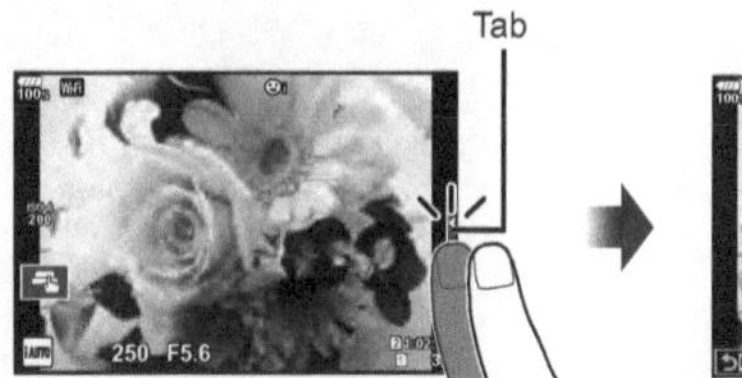

3 Use your finger to position the sliders.

- Tap OK to enter the setting.
- To cancel the live guide setting, tap MENU on the screen.
- When [Shooting Tips] is selected, select an item and tap OK to display a description.
- The effect of the selected level is visible in the display.
  If [Blur Background] or [Express Motions] is selected, the display will return to normal, but the selected effect will be visible in the final photograph.

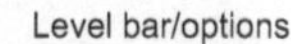

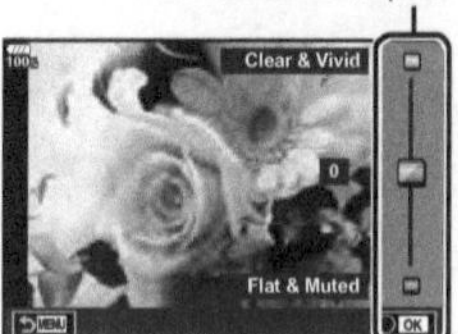

4 To set multiple live guides, repeat Steps 2 and 3.

- A check is displayed on the guide item for live guides that are already set.

5 Shoot.

- To clear the live guides from the display, press the **MENU** button.

- [Blur Background] and [Express Motions] cannot be set at the same time.
- If [RAW] is currently selected for image quality, image quality will automatically be set to [LN+RAW].
- Live guide settings are not applied to the RAW copy.
- Pictures may appear grainy at some live guide setting levels.
- Changes to live guide setting levels may not be visible in the monitor.
- Frame rates drop when [Blurred Motion] is selected.
- The flash cannot be used when a live guide is set.
- Choosing live guide settings that exceed the limits of the camera exposure meters may result in pictures that are overexposed or underexposed.

Using art filters

Using art filters, you can easily have fun with artistic effects.

■ Types of art filters

Pop Art I/II *	Creates an image that emphasizes the beauty of color.
Soft Focus	Creates an image that expresses a soft tone atmosphere.
Pale&Light Color I/II *	Creates an image that expresses warm light by scattering overall light and slightly overexposing the image.
Light Tone	Creates a high-quality image by softening both shadows and highlights.
Grainy Film I/II *	Creates an image that expresses the roughness of black and white images.
Pin Hole I/II/III *	Creates an image that looks as if it was taken using an old or toy camera by dimming the image perimeter.
Diorama I/II *	Creates a miniature-like image by emphasizing saturation and contrast, and blurring unfocused areas.
Cross Process I/II *	Creates an image that expresses a surreal atmosphere. Cross Process II creates an image that emphasizes magenta.
Gentle Sepia	Creates a high-quality image by drawing out shadows and softening the overall image.
Dramatic Tone I/II *	Creates an image that emphasizes the difference between brightness and darkness by partially increasing contrast.
Key Line I/II *	Creates an image that emphasizes edges and adds an illustrative style.
Watercolor I/II *	Creates a soft, bright image by removing dark areas, blending pale colors on a white canvas, and further softening contours.
Vintage I/II/III *	Expresses an everyday shot in a nostalgic, vintage tone using printed film discoloration and fading.
Partial Color I/II/III *	Impressively expresses a subject by extracting colors you want to emphasize and keeping everything else monotone.
ART BKT (ART bracketing)	Records images using all art filter options with a single shot. Press the **INFO** button on the selection screen to select a filter to record with.

* II and III are altered versions of the original (I).

1 Rotate the mode dial to **ART**.

- A menu of art filters will be displayed. Select a filter using the rear dial.
- Use △▽ to choose an effect. The effects available vary with the selected filter (Soft Focus Effect, Pin Hole Effect, Frame Effect, White Edge Effect, Star Light Effect, Color Filter, Monochrome Color, Blur Effect, or Shade Effect).
- Press the ⓞ button or press the shutter button halfway to select the highlighted item and exit the art filter menu.

2 Shoot.

- To choose a different setting, press the ⓞ button to display the art filter menu.

- To maximize the benefits of the art filters, some of the shooting function settings are disabled.
- If [RAW] is currently selected for image quality (P. 55, 88), image quality will automatically be set to [■N+RAW]. The art filter will be applied to the JPEG copy only.
- Depending on the subject, tone transitions may be ragged, the effect may be less noticeable, or the image may become more "grainy."
- Some effects may not be visible in live view or during movie recording.
- Playback may differ according to the filters, effects, or movie quality settings applied.

■ Using [Partial Color]

Record only selected hues in color.

1 Rotate the mode dial to **ART**.

2 Select [Partial Color].

3 Highlight a type or effect and press the ⓞ button.

- A color ring appears in the display.

4 Rotate the front dial or rear dial to select a color.

- The effect is visible in the display.

5 Shoot.

Custom Modes (C1, C2, C3)

Save settings to each of three Custom Modes for instant recall.

- Different settings can be saved to **C1**, **C2,** and **C3** using the [Reset / Custom Modes] (P. 87) in Shooting Menu 1.
- The settings for the selected Custom Mode will be recalled when you rotate the mode dial to **C1**, **C2**, or **C3**.

Shooting with touch screen operations

Tap to cycle through touch screen settings.

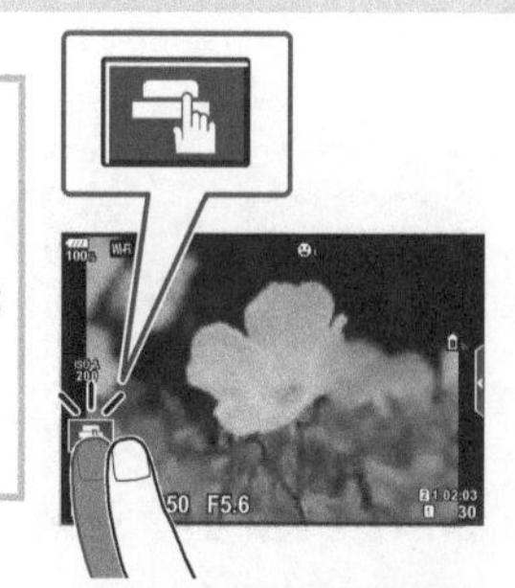

	Tap a subject to focus and automatically release the shutter. This function is not available in movie mode.
	Touch screen operations are disabled.
	Tap to display an AF target and focus on the subject in the selected area. You can use the touch screen to choose the position and size of the focus frame. Photographs can be taken by pressing the shutter button.

■ Previewing the subject ( )

1 Tap the subject in the display.

- An AF target will be displayed.
- Use the slider to choose the size of the frame.
- Tap Off to turn off the display of the AF target.

2 Use the slider to adjust the size of the AF target, and then tap to zoom in at the frame position.

- Use your finger to scroll the display when the picture is zoomed in.
- Tap 1x to cancel the zoom display.

- The situations in which touch screen operations are not available include the following.
  During multiple exposure, on the one-touch capture white balance screen, when buttons or dials are in use
- Do not touch the display with your fingernails or other sharp objects.
- Gloves or monitor covers may interfere with touch screen operation.
- You can disable the touch screen operation. ☞ [Touchscreen Settings] (P. 122)

Recording movies

Use the ◉ button to record movies.

1 Press the ◉ button to begin recording.

- Movie you are recording will be displayed on the monitor.
- If you put your eye to the viewfinder, movie you are recording will be displayed in the viewfinder.
- You can change the focus location by tapping the screen while recording.

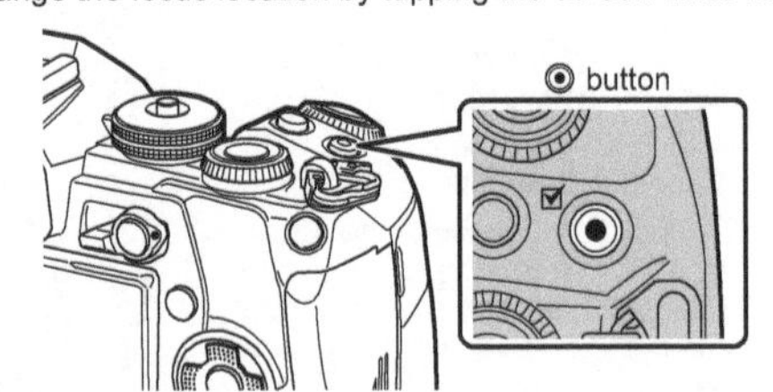

2 Press the ◉ button again to end recording.

- When using a camera with a CMOS image sensor, moving objects may appear distorted due to the rolling shutter phenomenon. This is a physical phenomenon whereby distortion occurs in the filmed image when shooting a fast-moving subject or due to camera shaking. In particular, this phenomenon becomes more noticeable when using a long focal length.
- If the size of the movie file being recorded exceeds 4 GB, the file will be split automatically. (Depending on shooting conditions, movies under 4 GB in size may be split into multiple files.)
- When recording movies, use an SD card that supports an SD speed class of 10 or higher.
- A UHS-II or UHS-I card with a UHS speed class of 3 or better is required when a movie resolution of [4K] or [C4K] or a bit rate of [A-I] (All-Intra) is selected in the [🎥◀⁝] menu.
- Choose an option other than [multi echo icon] (multi echo) for [Movie Effect] (P. 37) when recording at a movie resolution of [4K] or [C4K].
- If the camera is used for extended periods, the temperature of the image pickup device will rise and noise and colored fog may appear in images. Turn off the camera for a short time. Noise and colored fog may also appear in images recorded at high ISO sensitivity settings. If the temperature rises further, the camera will turn off automatically.
- When using a Four Thirds system lens, the AF will not operate while recording movies.
- The ◉ button cannot be used to record movies in the following instances:
  During multiple exposure (still photography also ends.), while shutter button is pressed halfway, during bulb/time/composite photography, during sequential shooting, during time lapse shooting

Using movie mode (🎥)

With movie mode (🎥), you can create movies that take advantage of the effects available in still photography mode. You can also apply an after-image effect or zoom in on an area of the image during movie recording.

■ Adding effects to a movie [Movie Effect]

You must first select [Movie Effect] on the screen that can be displayed by selecting the Video Menu > [🎥 Display Settings] > [🎥 Info Settings] > [Custom1] and pressing ▷ (P. 101).

1 Rotate the mode dial to 🎥.

2 Press the ◉ button to start shooting.

- Press the ◉ button again to end shooting.

3 Tap the on-screen icon of the effect you wish to use.

	Art Fade	Films with the selected picture mode effect. The fade effect is applied to the transition between scenes.
	Old Film	Randomly applies damage and dust-like noise similar to old movies.
	Multi Echo	Applies an after-image effect. After-images will appear behind moving objects.
	One Shot Echo	Applies an after-image for a short time after you tap the icon. The after-image will disappear automatically after a while.
	Movie Tele-converter	Zooms in on an area of the image without using the lens zoom. The selected position of the image can be zoomed in even while the camera is kept fixed.

Art Fade

Tap the icon. The effect will be applied gradually when you tap the icon of a picture mode.

Old Film

Tap the icon for the effect to be applied. Tap again to cancel the effect.

Multi Echo

Tap the icon for the effect to be applied. Tap again to cancel the effect.

One Shot Echo

Each tap of the icon adds the effect.

- Using the **INFO** button to change the information displayed in the monitor while data are being recorded cancels the selected [Movie Effect].

2 Shooting

Movie Tele-converter

1 Tap the icon to display the zoom frame.

- You can change the position of the zoom frame by tapping the screen or using △▽◁▷.
- Press and hold the ⓞ button to return the zoom frame to a central position.

2 Tap [icon] to zoom in on the area in the zoom frame.

- Tap [icon] to return to the zoom frame display.

3 Tap [□Off] or press the ⓞ button to cancel the zoom frame and exit Movie Tele-converter mode.

- The 2 effects cannot be applied simultaneously.
- Some effects may not be available depending on the picture mode.
- The sound of touch operations and button operations may be recorded.
- Art Fade cannot be used when shooting clips.
- When shooting slow/quick motion movies, you cannot use other movie effects than Movie Tele-converter.
- The drive mode displayed in the movie mode is the setting for still image shooting. Still image shooting is not available in the movie mode.
- The frame rate may drop if an art filter or a movie effect is used when a large image size such as [4K] or [C4K] is set.
- Art filters are not available for slow/quick motion movies.
- Movie effects are not available for picture modes of movie and for slow/quick motion movies.
- Movie Tele-converter is not available when [C4K] or [4K] is set for the image size.

Using the silencing function when shooting a movie

You can prevent the camera from recording operating sounds that occur due to camera operations while shooting.

The following functions are available as touch operations.

- Electronic zoom*1, recording volume, aperture, shutter speed, exposure compensation, ISO sensitivity, headphone volume*2

  *1 Only available with power zoom lenses

  *2 Only available when using headphones

Silent shooting tab

Tap the silent shooting tab to display the function items. After tapping an item, tap the displayed arrows to select the settings.

- The options available vary with the shooting mode.

Using various settings

Controlling exposure (Exposure ±)

Rotate the front dial to choose exposure compensation. Choose positive ("+") values to make pictures brighter, negative ("–") values to make pictures darker. Exposure can be adjusted by ±5.0 EV.

Negative (–) No compensation (0) Positive (+)

- Exposure compensation is not available in iAUTO.
- The viewfinder and live view display can only be changed up to ±3.0 EV. If the exposure exceeds ±3.0 EV, the exposure bar will begin flashing.
- Movies can be corrected in a range up to ±3.0 EV.

Selecting the AF target mode (AF target settings)

You can change the target selection method and target size. You can also select Face priority AF (P. 40).

1 Press the **Fn1** button to display the AF target.

- The AF target can also be displayed by pressing the arrow pad.

2 Use the front dial during AF target selection to choose a selection method.

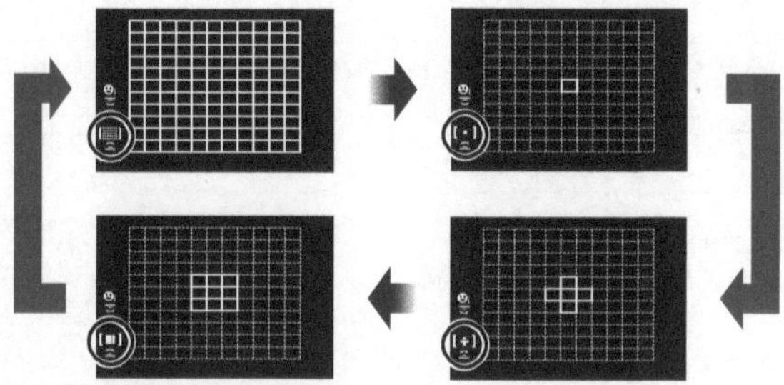

[▦] (All Targets)	The camera automatically chooses from the full set of focus targets.
[▪] (Single Target)	You can choose a single AF target.
[⁘] (5-Target Group)	The camera automatically chooses from the targets in the selected five-target group.
[⁙] (9-Target Group)	The camera automatically chooses from the targets in the selected nine-target group.

- The single target mode is automatically applied in movie shooting if the group target mode is set.

Setting the AF target

Select the single target or group target position.

1 Press the **Fn1** button to display the AF target.

- The AF target can also be displayed by pressing the arrow pad.

2 Use the arrow pad to position the AF target during AF target selection.

- The size and number of the AF target changes depending on the [Digital Tele-converter] (P. 88), [Image Aspect] (P. 54), and group target (P. 39) settings.
- Use the [[·⁞·] Custom Settings] (P. 112) in the custom menu to choose the roles of the dials and △▽◁▷ during AF target selection.

Face priority AF/Eye priority AF

The camera detects faces and adjusts focus and digital ESP.

1 Press the **Fn1** button to display the AF target.

- The AF target can also be displayed by pressing the arrow pad.

2 Use the rear dial to select an option during AF target selection.

Selected option

☺	**Face Priority On**	Face priority is on.
☺OFF	**Face Priority Off**	Face priority is off.
☺	**Face & Eye Priority On**	The autofocus system selects the eye closest to the camera for face-priority AF.
☺R	**Face & R. Eye Priority On**	The autofocus system selects the eye on the right for face-priority AF.
☺L	**Face & L. Eye Priority On**	The autofocus system selects the eye on the left for face-priority AF.

4 Point the camera at your subject.

- If a face is detected, it will be indicated by a white frame.

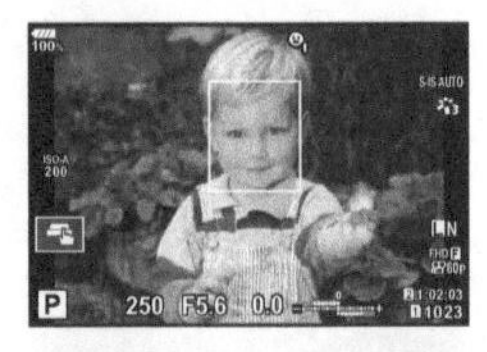

5 Press the shutter button halfway to focus.

- When the camera focuses on the face in the white frame, the frame will turn green.
- If the camera is able to detect the subject's eyes, it will display a green frame over the selected eye. (eye priority AF)

6 Press the shutter button the rest of the way down to shoot.

- Depending on the subject and the art filter setting, the camera may not be able to correctly detect the face.

- When set to [ESP] (Digital ESP metering)] (P. 45, 51), metering is performed with priority given to faces.
- Face priority is also available in [MF] (P. 43, 51). Faces detected by the camera are indicated by white frames.

Zoom frame AF/Zoom AF (Super Spot AF)

You can zoom in on a portion of the frame when adjusting focus. Choosing a high zoom ratio allows you to use autofocus to focus on a smaller area than is normally covered by the AF target. You can also position the focus target more precisely.

- To use Super Spot AF, you must first assign [Q] to a button with Button Function (P. 66).

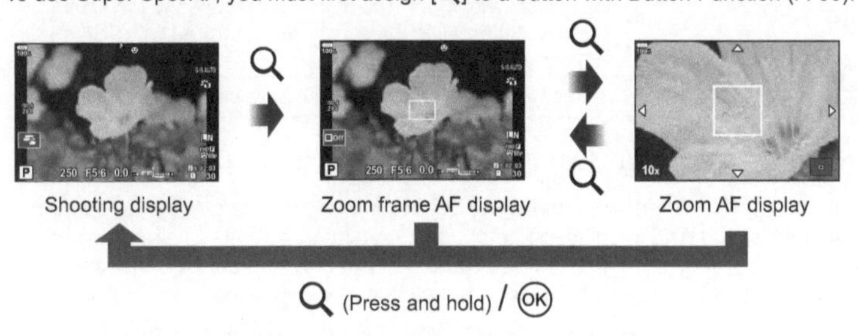

1 Press and release the Q button to display the zoom frame.

- If the subject has been focused using autofocus immediately before the button is pressed, the zoom frame will be displayed at the current focus position.
- Use △▽◁▷ to position the zoom frame.
- Press the **INFO** button and use △▽ to choose the zoom ratio. (×3, ×5, ×7, ×10, ×14)

2 Press and release the Q button again to zoom in on the zoom frame.

- Use △▽◁▷ to position the zoom frame.
- You can change the zoom ratio by rotating the front dial (front dial) or rear dial (rear dial).

3 Press the shutter button halfway to initiate autofocus.

- The camera will focus using the subject in the frame at the center of the screen. Use △▽◁▷ to choose a different focus position.

- Zoom is visible only in the monitor and has no effect on the resulting photographs.
- While zoomed in, a sound is made by the IS (Image Stabilizer).

Changing ISO sensitivity (ISO)

Increasing ISO sensitivity increases noise (graininess) but allows photographs to be taken when lighting is poor. The setting recommended in most situations is [AUTO], which starts at ISO 200—a value that balances noise and dynamic range—and then adjusts ISO sensitivity according to shooting conditions.

1 Set the **Fn** lever to the position 2, and rotate the front dial to select a value.

- Exposure compensation can be adjusted by rotating the front dial in **M** mode.

AUTO	The sensitivity is set automatically according to the shooting conditions. The upper limit of ISO sensitivity and the shutter speed to start raising the sensitivity can be set with [ISO-Auto Set] in Custom menu (P. 117).
LOW, 200–25600	The sensitivity is set to the selected value.

Adjusting color (WB (white balance))

White balance (WB) ensures that white objects in images recorded by the camera appear white. [AUTO] is suitable in most circumstances, but other values can be selected according to the light source when [AUTO] fails to produce the desired results or you wish to introduce a deliberate color cast into your images.

1 Set the **Fn** lever to the position 2, and rotate the rear dial to select a value.

- ISO sensitivity can be adjusted by rotating the rear dial in **M** mode.

WB mode		Color temperature	Light conditions
Auto white balance	**AUTO**	—	For most light conditions (when there is a white portion framed on the monitor). Use this mode for general use.
Preset white balance		5300 K	For shooting outdoors on a clear day, or to capture the reds in a sunset or the colors in a fireworks display
		7500 K	For shooting outdoors in the shadows on a clear day
		6000 K	For shooting outdoors on a cloudy day
		3000 K	For shooting under a tungsten light
		4000 K	For shooting under a fluorescent light
		—	For underwater photography
	WB⚡	5500 K	For flash shooting
One-touch white balance	1/ 2/ 3/ 4	Color temperature set by one-touch WB	Press the **INFO** button to measure white balance using a white or gray target when using a flash or other light source of an unknown type or when shooting under mixed lighting. ☞ "One-touch white balance" (P. 43)
Custom white balance	**CWB**	2000 K–14000 K	After pressing the **INFO** button, use ◁▷ to select a color temperature and then press the ⓞ button.

One-touch white balance

Measure white balance by framing a piece of paper or other white object under the lighting that will be used in the final photograph. This is useful when shooting a subject under natural light, as well as under various light sources with different color temperatures.

1 Select [1], [2], [3], or [4] (one-touch white balance 1, 2, 3, or 4) and press the **INFO** button.

2 Photograph a piece of colorless (white or gray) paper.
- Frame the paper so that it fills the display and no shadows fall it.
- The one-touch white balance screen appears.

3 Select [Yes] and press the ⓞ button.
- The new value is saved as a preset white balance option.
- The new value is stored until one-touch white balance is measured again. Turning the power off does not erase the data.

Choosing a focus mode (AF Mode)

You can select a focusing method (focus mode). You can choose separate focusing methods for still photography mode and movie mode.

1 Press the **AF** [◉] button.

2 Rotate the rear dial to select a value.

S-AF (Single AF)	The camera focuses once when the shutter button is pressed halfway. When the focus is locked, a beep sounds, and the AF confirmation mark and the AF target mark light up. This mode is suitable for taking pictures of still subjects or subjects with limited movement.
C-AF (Continuous AF)	The camera repeats focusing while the shutter button remains pressed halfway. When the subject is in focus, the AF confirmation mark lights up on the monitor and the beep sounds when the focus is locked at the first time. Even if the subject moves or you change the composition of the picture, the camera continues trying to focus.
MF (Manual focus)	This function allows you to manually focus on any position by operating the focus ring on the lens. Near ∞ Focus ring
S-AF+MF (Simultaneous use of S-AF mode and MF mode)	After pressing the shutter button halfway to focus in [S-AF] mode, you can turn the focus ring to fine-adjust focus manually.

C-AF+TR (AF tracking)	Press the shutter button halfway to focus; the camera then tracks and maintains focus on the current subject while the shutter button is held in this position. • The AF target is displayed in red if the camera can no longer track the subject. Release the shutter button and then frame the subject again and press the shutter button halfway. • Tracking range will be narrow when using a Four Thirds system lens. Autofocus does not work while the AF target is displayed in red even if the camera is tracking the subject.
PreMF (Preset MF)	The camera automatically focuses on the preset focus point when shooting.

- The camera may be unable to focus if the subject is poorly lit, obscured by mist or smoke, or lacks contrast.
- When using a Four Thirds system lens, AF will not be available during movie recording.
- AF-mode selection is not available if the lens MF clutch is set to the MF position and [Operative] is selected for [MF Clutch] (P. 112) in Custom Menu.

Setting a focus position for PreMF

1 Select [Preset MF] for AF mode.

2 Press the **INFO** button.

3 Press the shutter button halfway to focus.

- Focus can be adjusted by rotating the focus ring.

4 Press the ⓞⓚ button.

- The PreMF function can be recalled by pressing the button to which [Preset MF] is assigned in [📷 Button Function]. Press the button again to return to the original AF.
- The distance for the preset focus point can be set with [Preset MF distance] in Custom Menu (P. 112).

Choosing how the camera measures brightness (Metering)

You can choose how the camera meters subject brightness.

1 Press the **AF** [•] button.

2 Rotate the front dial to select a value.

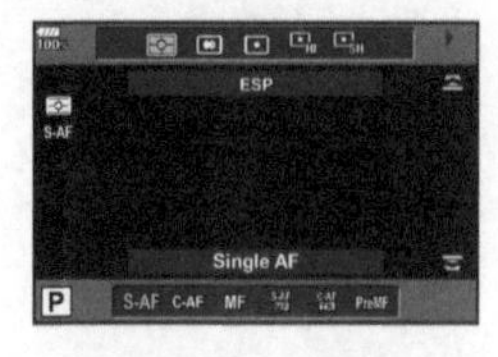

ESP	**Digital ESP metering**	Meters exposure in 324 areas of the frame and optimizes exposure for the current scene or portrait subject (if an option other than [OFF] is selected for [☺ Face Priority]). This mode is recommended for general use.
[(•)]	**Center weighted averaging metering**	Provides the average metering between the subject and the background lighting, placing more weight on the subject at the center.
[•]	**Spot metering**	Meters a small area (about 2% of the frame) with the camera pointed at the object you wish to meter. Exposure will be adjusted according to the brightness at the metered point.
[•]HI	**Spot metering (highlight)**	Increases spot metering exposure. Ensures bright subjects appear bright.
[•]SH	**Spot metering (shadow)**	Lowers spot metering exposure. Ensures dark subjects appear dark.

- The spot metering position can be set at the selected AF target (P. 118).

Locking the exposure (AE Lock)

You can lock the exposure by pressing the **AEL/AFL** button. Use this when you want to adjust the focus and exposure separately or when you want to shoot several images at the same exposure.

- If you press the **AEL/AFL** button once, the exposure is locked and AEL is displayed. ☞ "AEL/AFL" (P. 123)
- Press the **AEL/AFL** button once again to release the AE Lock.
- The lock will be released if you operate the mode dial, **MENU** button, or ⓞ button.

Performing the sequential/self-timer shooting

You can take a series of pictures by keeping the shutter button pressed all the way down. Alternatively, you can take pictures using the self-timer.

1 Press the ⎙HDR button.

2 Rotate the rear dial to select a value.

Single	1 frame is taken at a time when the shutter button is pressed (normal shooting mode, single-frame shooting).
Sequential High	Pictures are taken at up to about 15 frames per second (fps) while the shutter button is pressed all the way down. Focus, exposure, and white balance are fixed at the values for the first shot in each series.
Sequential Low	Pictures are taken at up to about 10 frames per second (fps) while the shutter button is pressed all the way down. Focus and exposure are fixed according to the options selected for [AF Mode] (P. 43, 51) and [AEL/AFL] (P. 123).
12s **12 sec**	Press the shutter button halfway to focus, the rest of the way down to start the timer. First, the self-timer lamp lights up for approximately 10 seconds, then it blinks for approximately 2 seconds and the picture is taken.
2s **2 sec**	Press the shutter button halfway to focus, the rest of the way down to start the timer. The self-timer lamp blinks for approximately 2 seconds, and then the picture is taken.
C **Custom Self-timer**	Press the **INFO** button to set [Timer], [Number of Frames], [Interval Length], and [Every Frame AF]. Select a setting using ◁▷, and adjust the setting by using the rear dial (). If [Every Frame AF] is set to [On], each frame is automatically focused before shooting.
Anti-Shock	The miniscule camera shaking caused by shutter movements can be reduced during continuous shooting and self-timer shooting (P. 47).
Silent	The sound of the shutter can be muted during sequential shooting and self-timer shooting (P. 47).
Pro Capture High	Sequential shooting begins when you press the shutter button halfway. Press the shutter button all the way down to begin recording captured images to the card, including those for a halfway press (P. 48). Focus, exposure, and white balance are fixed at the values for the first shot in each series.
Pro Capture Low	Sequential shooting begins when you press the shutter button halfway. Press the shutter button all the way down to begin recording captured images to the card, including those for a halfway press (P. 48). Focus and exposure are fixed according to the options selected for [AF Mode] (P. 43, 51) and [AEL/AFL] (P. 123).
High Res Shot	Still pictures can be taken in a higher resolution (P. 48).

- To cancel the activated self-timer, press ∇.
- Fix the camera securely on a tripod for self-timer shooting.
- If you stand in front of the camera to press the shutter button when using the self-timer, the picture may be out of focus.
- When you are using ♢L or ProCapL, live view is displayed. When you are using ♢H or ProCapH, the image shot immediately before is displayed.
- The speed of sequential shooting varies depending on the lens you are using and the focus of the zoom lens.
- During sequential shooting, if the battery level icon blinks due to low battery, the camera stops shooting and starts saving the pictures you have taken on the card. The camera may not save all of the pictures depending on how much battery power remains.
- You can set unused functions not to be displayed in the options. ☞ [♢/⏲ Settings] (P. 115)
- The frame advance rate for sequential shooting drops when [ISO] is set to 8000 or higher. The maximum frame advance rate in silent and pro-capture modes is 30 fps.
- Photographs taken in silent and pro-capture modes may be distorted if the subject or camera moves quickly during shooting.

2

Shooting

Shooting without the vibration caused by shutter button operations (Anti-Shock [♦])

To prevent camera shake caused by the small vibrations that occur during shutter operations, shooting is performed using an electronic front-curtain shutter.
This is used when shooting with a microscope or a super telephoto lens.
You must first set [Anti-Shock [♦]] in 📷2 Shooting Menu 2 to the settings other than [Off] (P. 98).

1 Press the ♢⏲HDR button.

2 Select one of the items marked ♦ using the rear dial and press the Ⓞ button.

3 Shoot.
   - When the set time has elapsed, the shutter is released and the picture is taken.

Shooting without shutter sound (Silent [♥])

In situations where the shutter sound is a problem, you can shoot without making a sound. Shooting is performed using electronic shutters for both the front and rear curtains so that the miniscule camera shaking caused by shutter movements can be reduced, just as in anti-shock shooting.
You can change the time between the shutter button is pressed all the way down and the shutter is released in [Silent [♥]] in 📷2 Shooting Menu 2. Set to [Off] to hide this setting item (P. 98).

1 Press the ♢⏲HDR button.

2 Select one of the items marked ♥ using the rear dial and press the Ⓞ button.

3 Shoot.
   - When the shutter is released, the monitor screen will go dark for a moment. No shutter sound will be emitted.

- The desired results may not be achieved under flickering light sources such as fluorescent or LED lamps or if the subject moves abruptly during shooting.

Shooting without a release time lag (Pro Capture shooting)

To resolve the time lag from when you press the shutter button all the way down until the recording of images starts, the sequential shooting using the electronic shutter starts when the shutter button is pressed halfway and the recording of images including those for a halfway press to the card starts when the shutter button is pressed all the way down.

ProCapH is good for subjects with minor changes in shooting distance, and ProCapL is good for subjects with changes in shooting distance.

1 Press the ◻︎ HDR button.

2 Select ProCapH or ProCapL using the rear dial and press the OK button.

3 Press the shutter button halfway to start shooting.

4 Press the shutter button all the way down to start recording to the card.

- Selecting ProCapL restricts aperture to values between maximum aperture and f8.0.
- Pro Capture shooting is not available when connected to Wi-Fi.
- Pro capture is available only with OLYMPUS Micro Four Thirds lenses.
- The camera will continue shooting for up to a minute while the shutter button is pressed halfway. To resume shooting, press the button halfway again.
- Flickering occurred by fluorescent lights or the large movement of the subject, etc. may cause distortions in images.
- The monitor will not be blacked out and the shutter sound will not be emitted while bracketing.
- The slowest shutter speed is limited.
- The sequential shooting speed, the number of pre-captured images, and the shot limit can be set in [ProCap] of [◻︎L Settings] or [◻︎H Settings] (P. 114) in Custom Menu.
- The display refresh rate may drop below the value selected for [Frame Rate] (P. 115) in the custom menu depending on subject brightness and the options selected for ISO sensitivity and exposure compensation.
- Choosing ProCapH when [C-AF] or [C-AF+TR] is selected changes the autofocus mode (P. 43, 51) to [S-AF].

Shooting still pictures in a higher resolution (High Res Shot)

When shooting an unmoving subject, you can shoot pictures in a higher resolution. A high resolution image is recorded by shooting several times while moving the imaging sensor. Attach the camera to a tripod or other stabilizing item and then shoot.

You can change the time between the shutter button is pressed all the way down and the shutter is released in [High Res Shot] in ◻︎2 Shooting Menu 2. Set to [Off] to hide this setting item (P. 99).

After setting high resolution shooting, you can select the image quality of the high resolution shooting using image quality mode (P. 55, 88).

1 Press the ◻︎ HDR button.

2 Select ◻︎ using the rear dial and press the OK button.

3 Shoot.

- If the camera is unstable, ◻︎ will blink. Wait for the blinking to stop before shooting.
- Shooting is complete when the green ◻︎ (high resolution) icon clears from the display.

- Choose from JPEG (50M F or 25M F) and JPEG+RAW modes. When image quality is set to RAW+JPEG, the camera saves a single RAW image (extension ".ORI") before combining it with the high resolution shooting. Pre-combination RAW images can be played back using the latest version of OLYMPUS Viewer 3.
- Image quality may drop under flickering light sources such as fluorescent or LED lamps.
- [Image Stabilizer] (P. 53) is set to [Off].

Taking HDR (High Dynamic Range) images

The camera shoots several images and automatically combines them into an HDR image. You can also shoot several images and perform HDR imaging on a computer (HDR bracketing photography).
Exposure compensation is available with [HDR1] and [HDR2] in **P**, **A**, and **S** modes. In **M** mode, exposure can be adjusted as desired for HDR photography.

1 Press the **HDR** button.

2 Rotate the front dial to select a setting.

<table>
<tr><td>HDR1</td><td rowspan="2">Four shots are taken, each with a different exposure, and the shots are combined into one HDR image inside the camera. HDR2 provides a more impressive image than HDR1. ISO sensitivity is fixed to 200. Also, the slowest available shutter speed is 4 seconds and the longest available exposure is 15 seconds.</td></tr>
<tr><td>HDR2</td></tr>
<tr><td>3F 2.0EV</td><td rowspan="5">HDR bracketing is performed. Select the number of images and the exposure difference.
HDR imaging processing is not performed.</td></tr>
<tr><td>5F 2.0EV</td></tr>
<tr><td>7F 2.0EV</td></tr>
<tr><td>3F 3.0EV</td></tr>
<tr><td>5F 3.0EV</td></tr>
</table>

3 Shoot.
- When you press the shutter button, the camera automatically shoots the set number of images.

- If you shoot with a slower shutter speed, there may be more noticeable noise.
- Attach the camera to a tripod or other stabilizing item and then shoot.
- The image displayed on the monitor or in the viewfinder while shooting will differ from the HDR-processed image.
- In the case of [HDR1] or [HDR2], the HDR-processed image will be saved as a JPEG file. When the image quality mode is set to [RAW], the image is recorded in RAW+JPEG.
- In the case of [HDR1] or [HDR2], the picture mode is fixed to [Natural] and the color setting is fixed to [sRGB].
- Flash photography, bracketing, multiple exposure, and time lapse shooting cannot be used at the same time as HDR photography.

Setting in super control panel

Other main shooting functions can be set in the LV super control panel.
Press the OK button to display the LV super control panel.
Change settings using △▽◁▷ or touch operations.

LV super control panel

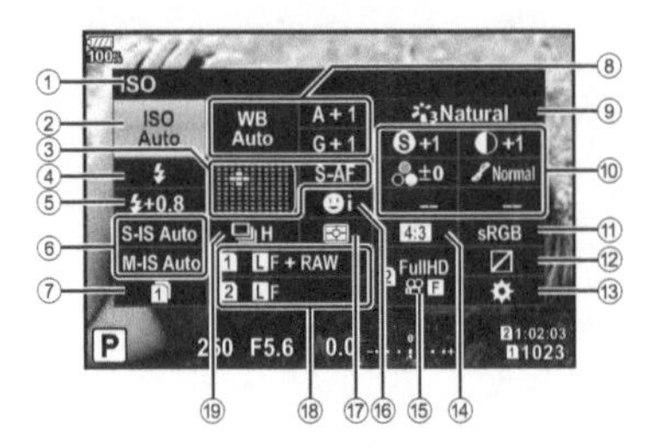

Super control panel

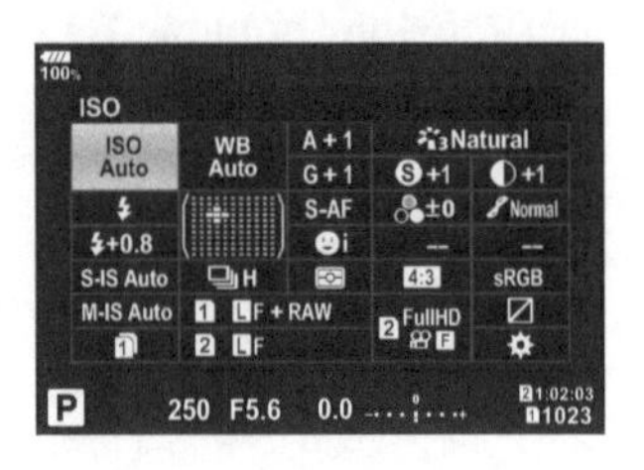

Settings that can be modified using LV super control panel

① Currently selected option
② ISO sensitivity ....P. 51
③ AF mode ....P. 51
AF target ....P. 40
④ Flash mode ....P. 57
⑤ Flash intensity control ....P. 60
⑥ Image stabilizer ....P. 53
⑦ Save Settings ....P. 54
⑧ White balance ....P. 52
White balance compensation ....P. 52
⑨ Picture mode ....P. 61
⑩ Sharpness ....P. 62
Contrast ....P. 62
Saturation ....P. 63
Gradation ....P. 63
Color filter ....P. 64
Monochrome color ....P. 64
Effect ....P. 65
Color*1 ....P. 34
Color/Vivid*2 ....P. 71
⑪ Color space ....P. 65
⑫ Highlight & shadow control ....P. 66
⑬ Button function assignment ....P. 66
⑭ Aspect ratio ....P. 54
⑮ ....P. 56
⑯ Face priority ....P. 40
⑰ Metering mode ....P. 51
⑱ ....P. 55
⑲ Sequential shooting/Self-timer ....P. 54

*1 Displayed when Partial Color is set.
*2 Displayed when Color Creator is set.

Changing ISO sensitivity (ISO)

You can set the ISO sensitivity.

☞ "Changing ISO sensitivity (ISO)" (P. 42)

1 Press the ⓞ button to display the LV super control panel.

2 Use △▽◁▷ to select [ISO].

3 Use the front dial to select an option.

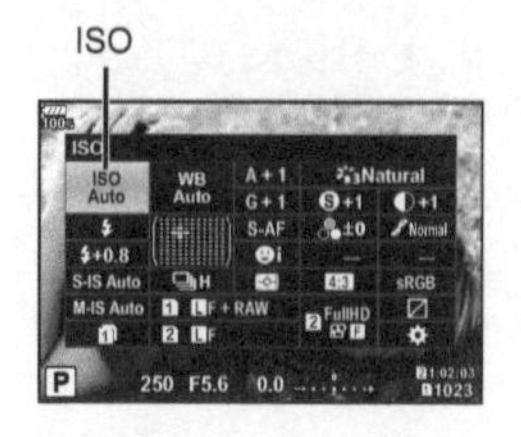

Choosing a focus mode (AF Mode)

You can select a focusing method (focus mode).

☞ "Choosing a focus mode (AF Mode)" (P. 43)

1 Press the ⓞ button to display the LV super control panel.

2 Use △▽◁▷ to select [AF Mode].

3 Use the front dial to select an option.

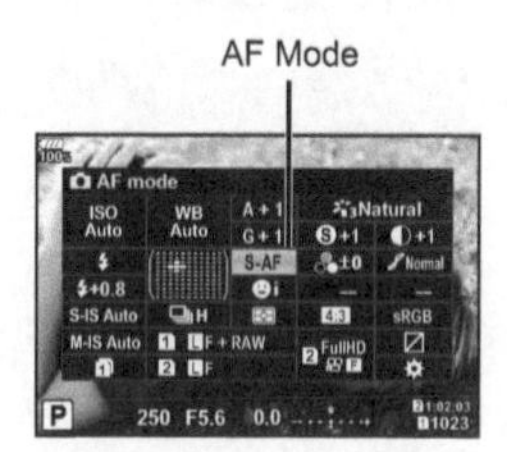

Choosing how the camera measures brightness (Metering)

You can choose how the camera meters subject brightness.

☞ "Choosing how the camera measures brightness (Metering)" (P. 45)

1 Press the ⓞ button to display the LV super control panel.

2 Use △▽◁▷ to select [Metering].

3 Use the front dial to select an option.

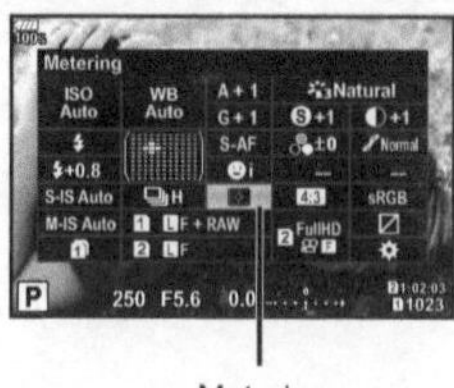

Adjusting color (WB (white balance))

You can set the white balance.

☞ "Adjusting color (WB (white balance))" (P. 42)

1. Press the ⓞ button to display the LV super control panel.
2. Use △▽◁▷ to select [WB].
3. Use the front dial to select an option.

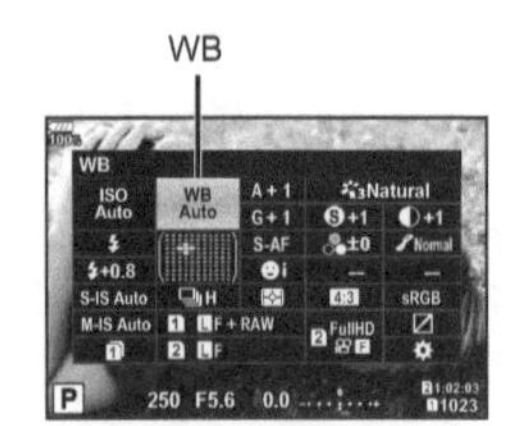

Making fine adjustments to white balance (WB Compensation)

You can set and finely adjust the compensation values for both auto white balance and preset white balance.

White balance compensation

1. Press the ⓞ button to display the LV super control panel.
2. Use △▽◁▷ to select [WB].
3. Use the front dial to select an option.
4. Use △▽◁▷ to select the white balance compensation.
5. Change a flash compensation value using the front dial.

For compensation on the A axis (Red-Blue)

Move the bar in the + direction to emphasize red tones and in the – direction to emphasize blue tones.

For compensation on the G axis (Green-Magenta)

Move the bar in the + direction to emphasize green tones and in the – direction to emphasize magenta tones.

- To set the same white balance in all white balance modes, use [All WB±] (P. 119).

Reducing camera shake (Image Stabilizer)

You can reduce the amount of camera shake that can occur when shooting in low light situations or shooting with high magnification.
The image stabilizer starts when you press the shutter button halfway.

1 Press the Ⓞ button to display the LV super control panel.

2 Use △▽◁▷ to select the image stabilizer.

3 Use the front dial to select an option.

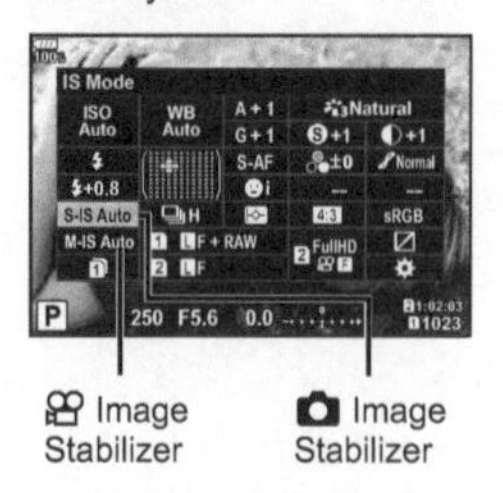

🎥 Image Stabilizer　　📷 Image Stabilizer

Still picture (S-IS)	OFF	Still-I.S. Off	Image stabilizer is off.
	S-IS AUTO	Auto I.S.	The camera detects the panning direction and applies the appropriate image stabilization.
	S-IS1	All Direction Shake I.S.	Image stabilizer is on.
	S-IS2	Vertical Shake I.S.	Image stabilization applies only to vertical (↕📷) camera shake.
	S-IS3	Horizontal Shake I.S.	Image stabilization applies only to horizontal (↔📷) camera shake. Use when panning the camera horizontally with the camera held in portrait orientation.
Movie (M-IS)	OFF	Movie-I.S. Off	Image stabilizer is off.
	M-IS1	All Direction Shake I.S.	The camera uses both sensor shift (VCM) and electronic correction.
	M-IS2	All Direction Shake I.S.	The camera uses sensor shift (VCM) correction only. Electronic correction is not used.

Using lenses other than Micro Four Thirds/Four Thirds System lenses

You can use focal length information to reduce camera shake when shooting with lenses that are not Micro Four Thirds or Four Thirds system lenses.

- Set [Image Stabilizer], press the Ⓞ button, press the **INFO** button, then use △▽◁▷ to select a focal length, and press the Ⓞ button.
- Choose a focal length between 0.1 mm and 1000.0 mm.
- Choose the value that matches the one printed on the lens.
- The image stabilizer cannot correct excessive camera shake or camera shake that occurs when the shutter speed is set to the slowest speed. In these cases, it is recommended that you use a tripod.
- When using a tripod, set [Image Stabilizer] to [OFF].
- When using a lens with an image stabilization function switch, priority is given to the lens side setting.
- When priority is given to the lens side image stabilization, [S-IS1] is used instead of [S-IS AUTO].
- You may notice an operating sound or vibration when the image stabilizer is activated.

Performing the sequential/self-timer shooting

You can take a series of pictures by keeping the shutter button pressed all the way down. Alternatively, you can take pictures using the self-timer.
☞ "Performing the sequential/self-timer shooting" (P. 46–48).

1 Press the ⓞ button to display the LV super control panel.

2 Use △▽◁▷ to select the sequential shooting/ self-timer.

3 Use the front dial to select an option.

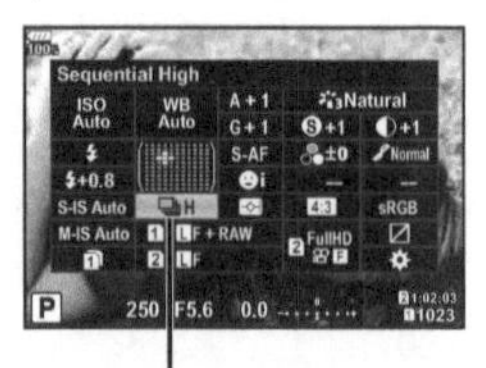
Sequential shooting/self-timer

Setting the image aspect

You can change the aspect ratio (horizontal-to-vertical ratio) when taking pictures. Depending on your preference, you can set the aspect ratio to [4:3] (standard), [16:9], [3:2], [1:1], or [3:4].

1 Press the ⓞ button to display the LV super control panel.

2 Use △▽◁▷ to select [Image Aspect].

3 Use the front dial to select an option.

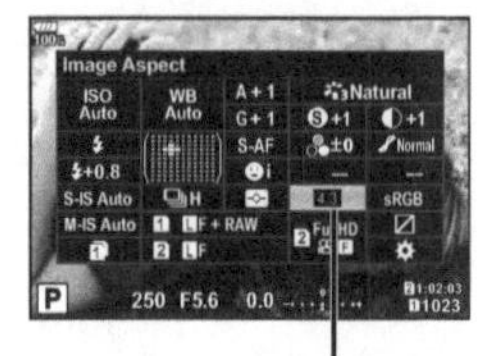
Image Aspect

- Image aspect can only be set for still images.
- JPEG images cropped to the selected aspect ratio are saved. RAW images are not cropped and are saved with the selected aspect ratio information.
- When RAW images are played back, the selected aspect ratio is shown by a frame.

Setting the saving method for shooting data (📷 Save Settings)

You can set how to record the shooting data to cards.

1 Press the ⓞ button to display the LV super control panel.

2 Use △▽◁▷ to select [📷 Save Settings].

3 Select an item using the front dial.

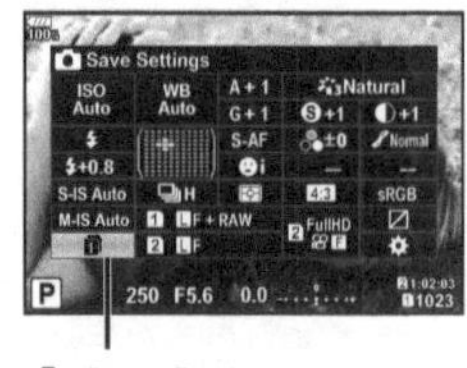
📷 Save Settings

(Standard)	This setting is applied if only one card with available space is in the camera. If two cards with available space are in the camera, images are recorded to the card specified in [📷 Save Slot] (P. 132).
(Auto Switch)	When the card specified in [📷 Save Slot] is full, recording switches to the other card (P. 132).

⇩□/□ (Dual Independent ↓□)	Images are recorded at the image quality mode specified for each card in slot 1 and 2 (P. 55, 88). Shooting will no longer be possible when either of the cards becomes full. Note that the image quality mode changes when [Dual independent ↓□] is selected; choose the desired mode before proceeding.
⇧□/□ (Dual Independent ↑□)	Images are recorded at the image quality mode specified for each card in slot 1 and 2 (P. 55, 88). When either of the cards becomes full, recording switches to the card with available space. Note that the image quality mode changes when [Dual independent ↑□] is selected; choose the desired mode before proceeding.
⇩□=□ (Dual Same ↓□)	Images are recorded at the same image quality mode for both cards. Shooting will no longer be possible when either of the cards becomes full.
⇧□=□ (Dual Same ↑□)	Images are recorded at the same image quality mode for both cards. When either of the cards becomes full, recording switches to the card with available space.

- If you press the ⓞ button, you can specify the card to record the shooting data to with [Card Slot Settings] of Custom Menu (P. 132).
- The image quality mode may change if you change the option selected for [📷 Save Settings] or replace a memory card with one that can contain a different amount of additional photographs. Check the image quality mode before taking photographs.

Selecting image quality (📷◀:·)

You can set an image quality mode for still images. Select a quality suitable for the application (such as for processing on PC, use on website etc.). This can be set for each card.

1 Press the ⓞ button to display the LV super control panel.

2 Use △▽◁▷ to select [📷◀:·].

- Image quality mode can be set for each card slot. If [📷 Save Settings] is set to [Dual Independent ↓□] or [DualIndependent ↑□] separate image quality modes can be set (P. 54).

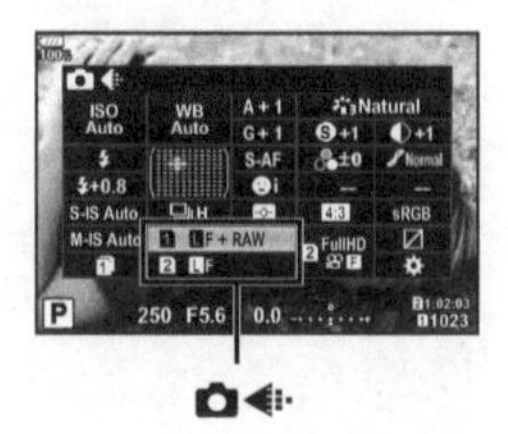

📷◀:·

3 Use the front dial to select an option.

- Choose from JPEG (LF, LN, MN, and SN) and RAW modes. Choose a JPEG+RAW option to record both a JPEG and a RAW image with each shot. JPEG modes combine image size (L, M, and S) and compression ratio (SF, F, N, and B).
- When you want to select a combination other than LF, LN, MN, and SN, change the [◀:· Set] (P. 119) settings in Custom Menu.
- During the high resolution shooting (P. 48), you can select between 50M F, 25M F, 50M F+RAW, and 25M F+RAW.
- Selecting [Dual independent ↓□] or [Dual independent ↑□] for [📷 Save Settings] changes the image quality mode; choose the desired mode before proceeding.
- The image quality mode may change if you change the option selected for [📷 Save Settings] or replace a memory card with one that can contain a different amount of additional photographs. Check the image quality mode before taking photographs.

RAW image data

This format (extension ".ORF") stores unprocessed image data for later processing. RAW image data cannot be viewed using other cameras or software, and RAW images cannot be selected for printing. JPEG copies of RAW images can be created using this camera. ☞ "Editing still images" (P. 105)

Selecting image quality (movie quality)

You can set a movie record mode suitable for the desired use.
Set the card recording image quality mode for the card set as the movie recording destination in [Card Slot Settings]. ☞ "Setting the card to record to" (P. 132)

1 Press the Ⓞ button to display the LV super control panel.

2 Use △▽◁▷ to select [movie quality].

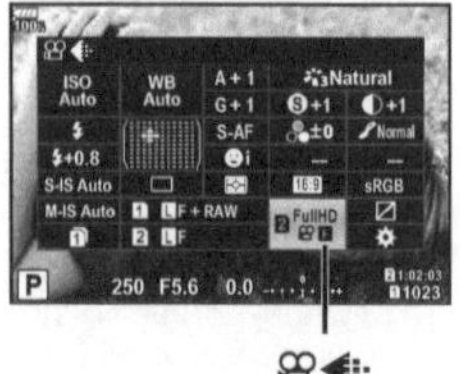

3 Use the front dial to select an option.

- To change movie record mode settings, press Ⓞ followed by the **INFO** button and rotate the rear dial.

Record mode	Application	Settings you can change
FHD F 30p (Full HD Fine 30p)*1	Shooting clips (P. 72)	Movie resolution, bit rate, frame rate, recording time*4
4K 30p (4K 30p)*1	Setting 1	Movie resolution, bit rate, frame rate*4
FHD SF 60p (Full HD Super Fine 60p)*1*2	Setting 2	Movie resolution, bit rate, frame rate*4
FHD F 60p (Full HD Fine 60p)*1*2	Setting 3	Movie resolution, bit rate, frame rate*4
FHD N 60p (Full HD Normal 60p)*1*2	Setting 4	Movie resolution, bit rate, frame rate*4
C4K 24p (C4K 24p)*1	Custom	Movie resolution, bit rate, frame rate, maximum clip recording time, slow/quick motion shooting*4
HD (1280×720, Motion JPEG)*3	Computer playback or editing	—

*1 File format: MPEG-4 AVC/H.264. The maximum size of individual file is limited to 4 GB. The maximum recording time of individual movie is limited to 29 minutes.

*2 All-Intra refers to a movie recorded without inter-frame compression. Movies in this format are suitable for editing but have a larger data size.

*3 The maximum size of individual file is limited to 2 GB.

*4 The record mode can be set.
☞ [🎥 Specification Settings] (P. 100)

- Depending on the type of card used, recording may end before the maximum length is reached.
- Bit rate selection may be unavailable at some settings.

Using a flash (Flash photography)

1 Remove the hot shoe cover and attach the flash to the camera.

- Slide the flash unit all the way in until it contacts the back of the shoe and is securely in place.

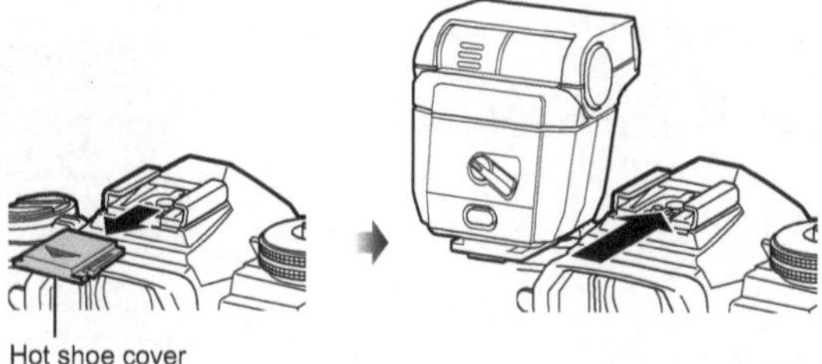

Changing orientation of the flash

You can change the vertical and horizontal orientation of the flash unit. Bounce shooting is also possible.

- Note that when used for bounce photography the flash may not fully illuminate the subject.

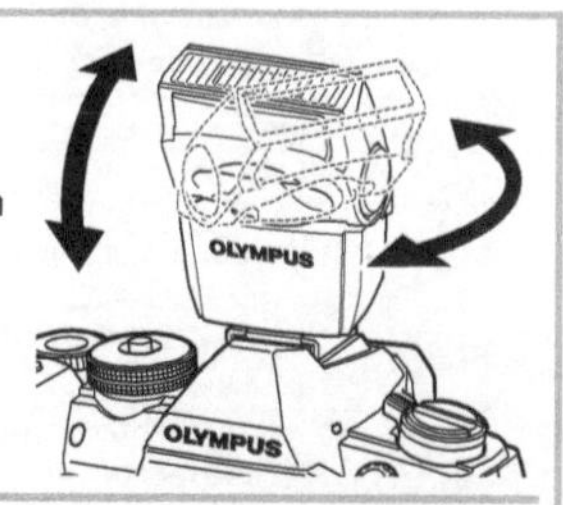

Removing the flash unit

Remove the flash unit while pressing the UNLOCK switch.

2 Set the flash **ON/OFF** lever to the ON position, and turn on the camera.

- When not using the flash, return the lever to the OFF position.

3 Press the Ⓞ button to display the LV super control panel.

4 Use △▽◁▷ to select [Flash Mode].

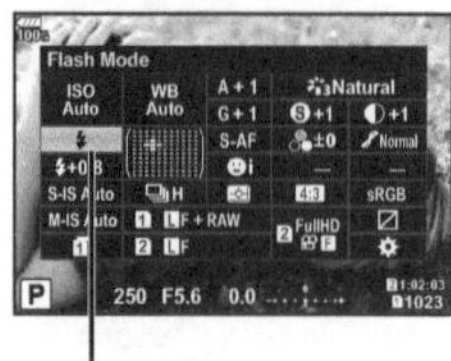

Flash Mode

5 Use the front dial to select an option.

- The options available and the order in which they are displayed vary depending on the shooting mode. ☞ "Flash modes that can be set by shooting mode" (P. 59)

⚡	**Flash**	The flash fires regardless of the light conditions.
ⓢ	**Flash off**	The flash does not fire.
⚡◉	**Red-eye reduction flash**	The flash fires so that the red-eye phenomenon is reduced.
⚡SLOW	**Slow synchronization (1st curtain)**	The flash fires with slow shutter speeds to brighten dimly-lit backgrounds.
◉SLOW	**Slow synchronization (1st curtain)/Red-eye reduction flash**	The slow synchronization is combined with the red-eye reduction flash.
⚡SLOW2	**Slow synchronization (2nd curtain)**	The flash fires immediately before the shutter closes to create trails of light behind moving light sources.
⚡FULL, ⚡1/4 etc.	**Manual flash**	For users who prefer manual operation. If you press the Ⓞ button followed by the **INFO** button, you can use the dial to adjust the flash level.

- In [⚡◉ (Red-eye reduction flash)], after the pre-flashes, it takes about 1 second before the shutter is released. Do not move the camera until shooting is complete.
- [⚡◉ (Red-eye reduction flash)] may not work effectively under some shooting conditions.
- When the flash fires, the shutter speed is set to 1/250 seconds or slower. When shooting a subject against a bright background with the flash, the background may be overexposed.
- The sync speed for silent mode and focus bracketing (P. 94) is 1/50 second. The sync speed at ISO sensitivities of 8000 and above and during ISO bracketing (P. 94) is 1/20 second.

Flash modes that can be set by shooting mode

Shooting mode	LV super control panel	Flash mode	Flash timing	Conditions for firing the flash	Shutter speed limit
P/A	⚡	Flash	1st curtain	Always fires	30 sec. – 1/250 sec.*
	⚡◉	Red-eye reduction			1/30 sec. – 1/250 sec.*
	⊛	Flash off	—	—	—
	◉ SLOW	Slow synchronization (red-eye reduction flash)	1st curtain	Always fires	60 sec. – 1/250 sec.*
	⚡SLOW	Slow Synchronization (1st curtain)			
	⚡ SLOW2	Slow synchronization (2nd curtain)	2nd curtain		
S/M	⚡	Flash	1st curtain	Always fires	60 sec. – 1/250 sec.*
	⚡◉	Red-eye reduction flash			
	⊛	Flash off	—	—	—
	⚡ SLOW2	Slow synchronization (2nd curtain)	2nd curtain	Always fires	60 sec. – 1/250 sec.*

- Only ⚡ and ⊛ can be set in iAUTO mode.

* The shutter speed is 1/250 seconds when using a separately sold external flash.

Minimum range

The lens may cast shadows over objects close to the camera, causing vignetting, or the flash may be too bright even at minimum output.

Lens	Approximate distance at which vignetting occurs
ED 12-40mm f2.8 PRO	0.6 m
ED 40-150mm f2.8 PRO	0.6 m

- External flash units can be used to prevent vignetting. To prevent photographs from being overexposed, select **A** or **M** mode, and choose a high aperture value or reduce ISO sensitivity.

Adjusting flash output (Flash intensity control)

Flash output can be adjusted if you find that your subject is overexposed, or is underexposed even though the exposure in the rest of the frame is appropriate.

1 Press the ⓞ button to display the LV super control panel.

2 Use △▽◁▷ to select [⚡☒].

3 Use the front dial to select an option.

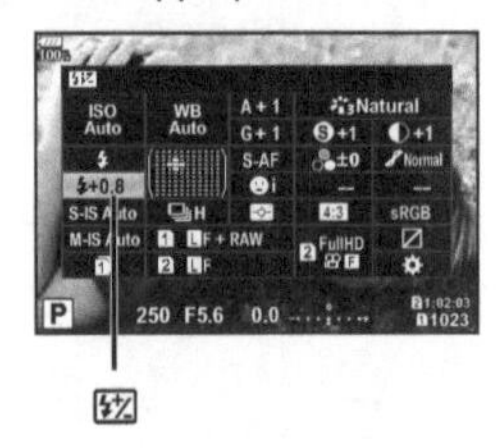

- This setting has no effect when the flash control mode for the external flash unit is set to MANUAL.
- Changes to flash intensity made with the external flash unit are added to those made with the camera.

Processing options (Picture Mode)

You can select a picture mode and make individual adjustments to contrast, sharpness, and other parameters (P. 62). Changes to each picture mode are stored separately.

■ Picture mode options

i-Enhance	Produces more impressive-looking results suited to the scene.
Vivid	Produces vivid colors.
Natural	Produces natural colors.
Muted	Produces flat tones.
Portrait	Produces beautiful skin tones.
Monochrome	Produces black and white tone.
Custom	Use to select one picture mode, set the parameters, and register the setting.
e-Portrait	Produces smooth skin textures. This mode cannot be used with bracket photography or when shooting movies.
Underwater	Produces vivid color finish perfect for underwater photos. • When [Underwater] is set, it is recommended that you set [+ WB] to [Off] (P. 119).
Color Creator	Provides a color finish set in Color Creator (P. 71).
ART 1 Pop Art	Uses Art Filter settings. Art effects can also be used.
ART 2 Soft Focus	
ART 3 Pale&Light Color	
ART 4 Light Tone	
ART 5 Grainy Film	
ART 6 Pin Hole	
ART 7 Diorama	
ART 8 Cross Process	
ART 9 Gentle Sepia	
ART 10 Dramatic Tone	
ART 11 Key Line	
ART 12 Watercolor	
ART 13 Vintage	
ART 14 Partial Color	

1 Press the ⓞ button to display the LV super control panel.

2 Use △▽◁▷ to select [Picture Mode].

3 Use the front dial to select an option.

- The items available for picture mode differ depending on the shooting mode (P. 24).
- You can set unused picture modes not to be displayed in the options. ☞ [Picture Mode Settings] (P. 115) Some picture modes cannot be hidden.

Picture Mode

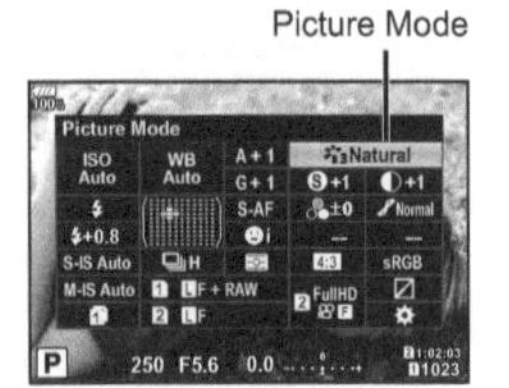

2 Shooting

Making fine adjustments to sharpness (Sharpness)

In the picture mode (P. 61) settings, you can make fine adjustments to sharpness and store the changes.

- Adjustments may not be available depending on the shooting mode (P. 24).

1 Press the ⓞ button to display the LV super control panel.

2 Use △▽◁▷to select [Sharpness].

3 Use the front dial to select an option.

Sharpness

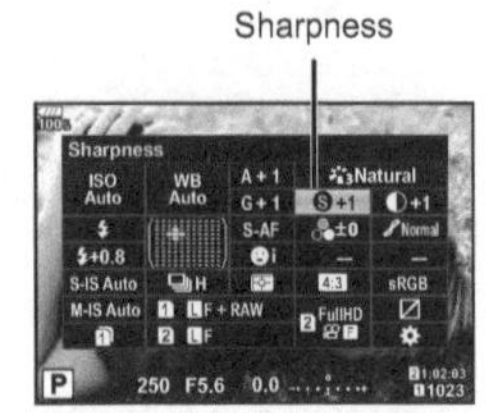

Making fine adjustments to contrast (Contrast)

In the picture mode (P. 61) settings, you can make fine adjustments to contrast and store the changes.

- Adjustments may not be available depending on the shooting mode (P. 24).

1 Press the ⓞ button to display the LV super control panel.

2 Use △▽◁▷to select [Contrast].

3 Use the front dial to select an option.

Contrast

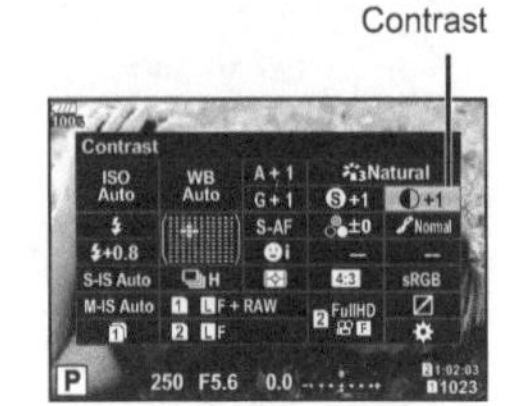

Making fine adjustments to saturation (Saturation)

In the picture mode (P. 61) settings, you can make fine adjustments to saturation and store the changes.

- Adjustments may not be available depending on the shooting mode (P. 24).

1 Press the ⓚ button to display the LV super control panel.

2 Use △▽◁▷ to select a [Saturation].

3 Use the front dial to select an option.

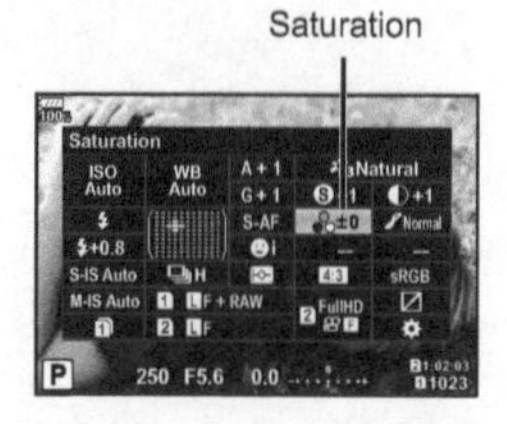

Making fine adjustments to tone (Gradation)

In the picture mode (P. 61) settings, you can make fine adjustments to tone and store the changes.

- Adjustments may not be available depending on the shooting mode (P. 24).

1 Press the ⓚ button to display the LV super control panel.

2 Use △▽◁▷ to [Gradation].

3 Use the front dial to select an option.

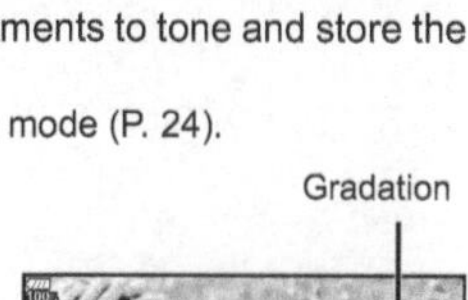

AUTO: Auto	Divides the image into detailed regions and adjusts the brightness separately for each region. This is effective for images with areas of large contrast in which the whites appear too bright or the blacks appear too dark.
NORM: Normal	Use the normal mode for general uses.
HIGH: High Key	Uses a tone suitable for a bright subject.
LOW: Low Key	Uses a tone suitable for a dark subject.

Applying filter effects to monochrome pictures (Color Filter)

In the monochrome setting of picture mode (P. 61) settings, you can add and store a filter effect in advance. This creates a monochrome image in which the color matching the filter color is brightened and the complementary color is darkened.

Color Filter

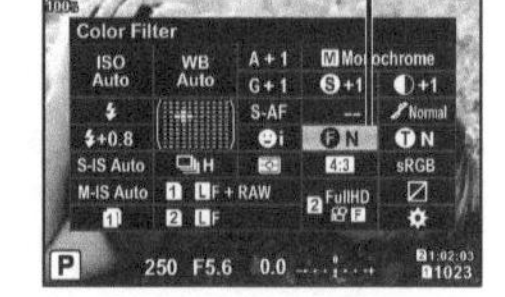

1 Press the ⓞⓚ button to display the LV super control panel.

2 Use △▽◁▷ to select [Picture Mode].

3 Select [Monochrome] using the front dial.

4 Use △▽◁▷ to select [Color Filter].

5 Select an item using the front dial.

N:None	Creates a normal black and white image.
Ye:Yellow	Reproduces clearly defined white cloud with natural blue sky.
Or:Orange	Slightly emphasizes colors in blue skies and sunsets.
R:Red	Strongly emphasizes colors in blue skies and brightness of crimson foliage.
G:Green	Strongly emphasizes colors in red lips and green leaves.

Adjusting the tone of a monochrome image (Monochrome Color)

In the monochrome setting of picture mode (P. 61) settings, you can add and store a color tint in advance.

Monochrome Color

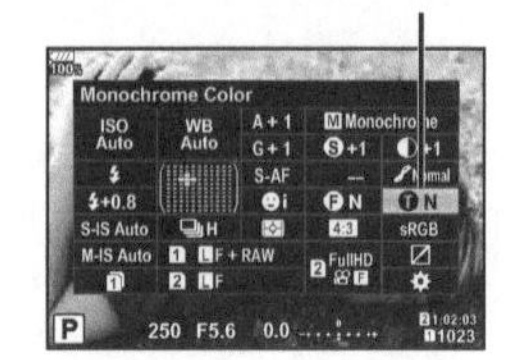

1 Press the ⓞⓚ button to display the LV super control panel.

2 Use △▽◁▷ to select [Picture Mode].

3 Select [Monochrome] using the front dial.

4 Use △▽◁▷ to select [Monochrome Color].

5 Use the front dial to select an option.

N:None	Creates a normal black and white image.
S:Sepia	Creates a sepia image.
B:Blue	Creates a bluish image.
P:Purple	Creates a purplish image.
G:Green	Creates a greenish image.

Adjusting i-Enhance effects (Effect)

You can set the i-Enhance effect strength in picture mode (P. 61).

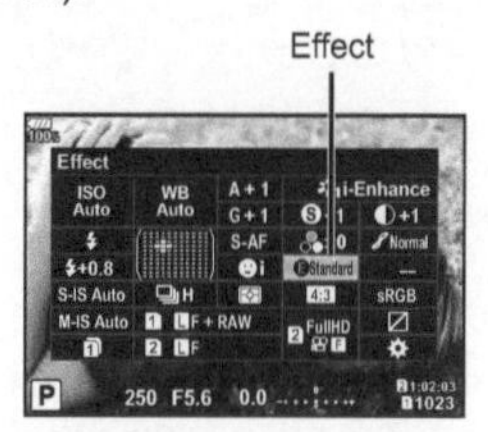

1 Press the ⓞ button to display the LV super control panel.

2 Use △▽◁▷ to select [Effect].

3 Use the front dial to select an option.

(Effect: Low)	Adds a low i-Enhance effect to images.
(Effect: Standard)	Adds an i-Enhance effect between "low" and "high" to images.
(Effect: High)	Adds a high i-Enhance effect to images.

Setting the color reproduction format (Color Space)

You can select a format to ensure that colors are correctly reproduced when shot images are regenerated on a monitor or using a printer. This option is equivalent to the [Color Space] (P. 119) in Custom Menu.

1 Press the ⓞ button to display the LV super control panel.

2 Use △▽◁▷ to select [Color Space].

3 Use the front dial to select an option.

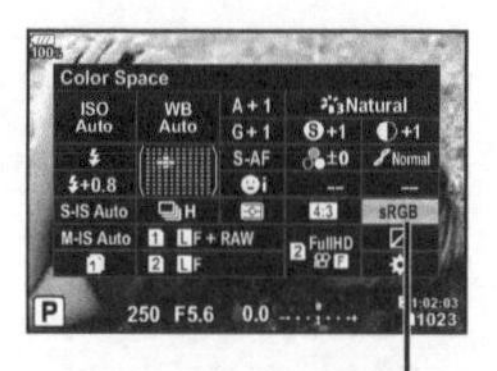

sRGB	This is the RGB color space standard stipulated by the International Electrotechnical Commission (IEC). Normally, use [sRGB] as the standard setting.
AdobeRGB	This is a standard provided by Adobe Systems. Compatible software and hardware such as a display, printer etc. are required for a correct output of images.

- [AdobeRGB] is not available in movie mode or with **ART** (P. 33) or HDR

Changing the brightness of highlights and shadows (Highlight&Shadow)

Use [Highlight&Shadow] to adjust the brightness of highlights and shadows.

1 Press the ⓞ button to display the LV super control panel.

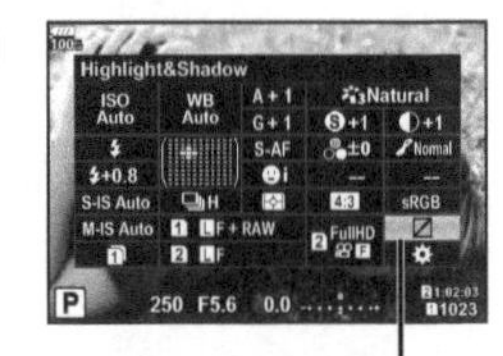

Highlight&Shadow

2 Use △▽◁▷ to select [Highlight&Shadow], then press the ⓞ button.

3 Use the rear dial to adjust shadows and the front dial to adjust highlights.

- Settings can be reset by pressing and holding the ⓞ button.
  Press the **INFO** button to view the mid-tone adjustment display.

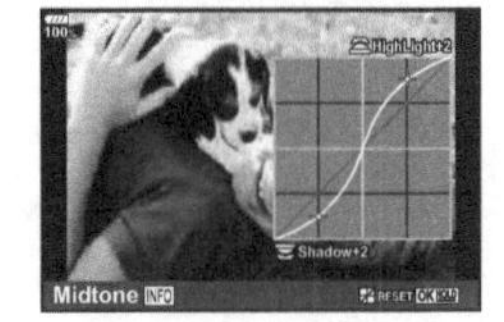

- The [Highlight&Shadow] can also be accessed via the button to which [Highlight&Shadow control] has been assigned using [Multi Function] option (P. 68).

Assigning functions to buttons (Button Function)

At default settings, the following functions are assigned to the buttons:

Button	Default
Fn1 Function	AF Area Select
Fn2 Function	Multi Function
◉ Function	◉ REC
AEL/AFL Function	AEL/AFL
Preview Function	Preview (Still picture), Peaking (Movie)
Magnify Function	Magnify (Still picture), Q (Movie)
IOI Function	IOI
◁▷ Function	AF Area Select (Still picture), Direct Function (Movie)
▷ Function	⚡ (Still picture), Electronic Zoom* (Movie)
▽ Function	▢ii/Ⓢ (still picture), ISO/WB (Movie)
B-Fn1 Function	AF Area Select
B-Fn2 Function	AEL/AFL
PBH✣ Function	AF Area Select (Still picture), Direct Function (Movie)
PBH▶ Function	⚡ (Still picture), Electronic Zoom* (Movie)
PBH▼ Function	▢ii/Ⓢ (still picture), ISO/WB (Movie)
L-Fn Function	AF Stop

* Only available with power zoom lenses

To change the function assigned to a button, follow the steps below.

1 Press the ⓞ button to display the LV super control panel.

2 Use △▽◁▷ to select [◘ Button Function] or [🎥 Button Function], then press the ⓞ button.
   - [◘ Button Function] of Custom Menu (P. 113) is displayed in still image shooting mode, and [🎥 Button Function] of [🎥 Button/Dial/Lever] (P. 100) is displayed in movie mode.

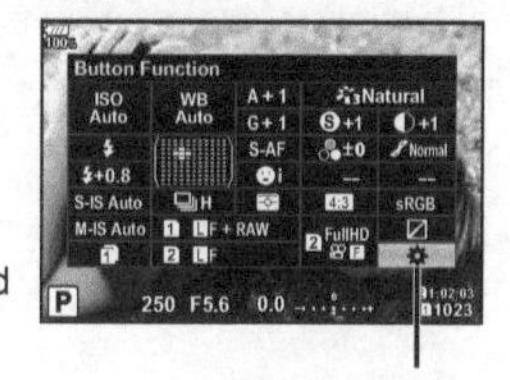

Button Function

3 Use the rear dial to select a button.

4 Rotate the front dial to switch the setting screen, then use the rear dial to select an option and press the ⓞ button.

- The options available vary from button to button.

AF Stop	Press the button to stop autofocus.
AEL/AFL (AEL/AFL)	Press the button to use AE lock or AF lock. The function changes according to the [AEL/AFL] (P. 123) setting. When AEL is selected, press the button once to lock the exposure and display AEL on the monitor. Press the button again to cancel the lock.
◉ REC	Press the button to record a movie.
(Preview)	Aperture is stopped down to the selected value while the button is pressed. If [On] is selected for [Lock] in [Settings] of Custom Menu (P. 116), the aperture is held at the selected value even when the button is released.
(One-touch white balance)	Press the shutter button while the button is pressed to acquire a white balance value (P. 43).
[·:·] (AF Area Select)	Press the button to choose the AF target (P. 39, 40).
[·:·] HP ([·:·] Home)	Press the button to select the AF home position saved with [[·:·] Set Home] (P. 112). Press the button again to return to the original position. If the camera is turned off when the home position is selected, the AF target position set before the home position is selected will be reset.
MF	Press the button to select [MF] mode. Press the button again to restore the previously selected AF mode. You can switch the focusing mode by rotating the dial while pressing the button.
RAW	Press the button to toggle between JPEG and RAW+JPEG image quality modes. You can switch the image quality mode by rotating the dial while pressing the button.
◘ TEST (Test Picture)	Press the shutter button while the button is pressed to display taken pictures without recording to the memory card.

/ (Underwater wide/ Underwater macro)	When using a waterproof protector, press the button to switch between and . Press and hold the button to return to the original shooting mode. When using a lens with an electronic zoom, switching between and is automatically set to the wide-angle setting and the telephoto setting.
(Exposure compensation)	Press the button to adjust exposure compensation. If you press the button while in **P**, **A**, or **S** modes, you can adjust exposure compensation with the dial or ◁▷. If you press the button while in **M** mode, you can change the shutter speed and aperture value with the dial or △▽◁▷.
(Digital Tele-converter)	Press the button to turn digital zoom [On] or [Off].
(Keystone Comp.)	Press the button once to display options for keystone compensation and again to save changes and exit. To resume normal photography, press and hold the button.
(Magnify)	Press the button to display the magnification frame and press it again to magnify the image. Press and hold the button to cancel magnified display.
HDR	Press the button to switch to HDR shooting with the saved settings. Press the button again to cancel HDR shooting. You can switch the HDR mode by rotating the dial while pressing the button.
BKT	Press the button to switch to BKT shooting with the saved settings. Press the button again to cancel BKT shooting. You can switch the BKT mode by rotating the dial while pressing the button.
ISO/WB	Press the button to adjust ISO sensitivity using the front dial and white balance using the rear dial.
WB/ISO	Press the button to adjust white balance using the front dial and ISO sensitivity using the rear dial.
Multi Function	To recall the selected multi function, press the button to which [Multi Function] has been assigned. ☞ "Using multi function options (Multi Function)" (P. 70)
Peaking	Press the button to turn on and off the peaking display. When peaking is displayed, histogram and highlight/ shadow display are not available. When using peaking, you can change colors and emphasis by pressing the **INFO** button.
Level Disp	Press the button to display the level gauge in the viewfinder and press it again to turn off the level gauge. The level gauge is available when [Style 1] or [Style 2] is set in [EVF Style] (P. 133).
|◯| (|◯| View Selection)	Press this button to turn on and off the live view. If the eye sensor is disabled, it switches between monitor display and EVF display.
(S-OVF)	Press the button to display an image in the viewfinder as an optical viewfinder image. will be displayed in the viewfinder. Press the button to end [S-OVF].

AF Limiter	Press the button to turn on and off the AF Limiter. You can switch the AF Limiter mode by rotating the dial while pressing the button.
Preset MF	Press the button to switch to Preset MF. Press the button again to return to the original AF setting. You can switch the focusing mode by rotating the dial while pressing the button.
Exif Lens (Lens Info Settings)	Press the button to display the lens information settings menu (P. 132).
IS Mode	Press the button to turn on and off the image stabilization. You can switch the image stabilization mode by rotating the dial while pressing the button.
ϟ (Flash Mode)	Press the button to choose a flash mode. You must first set [⇔ Function] to [Direct Function].
▭/◌	Press the button to choose a sequential shooting or self-timer option. You must first set [⇔ Function] to [Direct Function].
▣ (Switch ▣ Lock)	Press and hold the button to enable and disable touch screen operation. You must first set [⇔ Function] to [Direct Function].
Electronic Zoom	When using a lens with a power zoom function, after pressing the button, use the arrow pad for zooming operations. You must first set [⇔ Function] to [Direct Function].

- To use [▶ Function] and [▼ Function] options, you will first have to select [Direct Function] for [⇔ Function].
- To use [PBH▶ Function] and [PBH▼ Function] options, you will first have to select [Direct Function] for [PBH✥ Function].
- The [Direct Function] option for ⇔ button applies to each of △▽◁▷.
- Assign [·:·] to the ⇔ button to use it for AF target selection.
- The L-Fn button can be used for the functions available on some lenses.
- The multi-function button can be assigned the following roles:
  ▣ (Highlight&Shadow Control), ◎ (Color Creator), ISO/WB (ISO/WB), WB/ISO (WB/ISO), Q (Magnify), ▭ (Image Aspect), S-OVF (S-OVF), PEAK (Peaking)

■ Using multi function options (Multi Function)

Multiple functions can be assigned to a button by setting [Multi Function] to the button in [📷 Button Function] or [🎥 Button Function]. At default settings, multi function is assigned to the **Fn2** button.

Choosing a function

2 Shooting

1 Press and hold the **Fn2** button and rotate the front or rear dial.

- The menus are displayed.

2 Keep rotating the dial to choose a function.

- Release the button when the desired function is selected.

Using the selected function

Press the **Fn2** button. The function setting screen will be displayed.

When [🔍] is assigned to the **Fn2** button

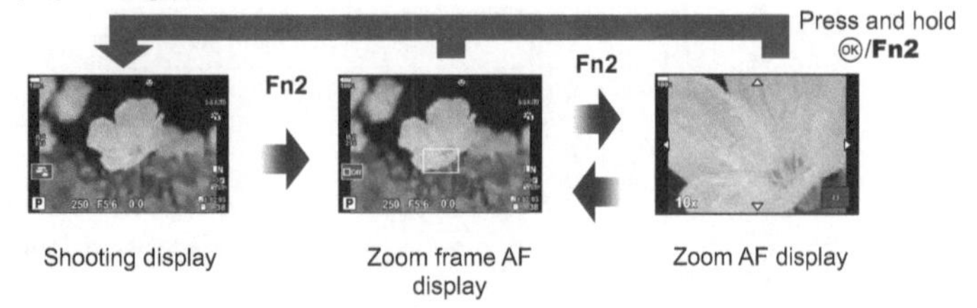

Shooting display

Zoom frame AF display

Zoom AF display

When other options are assigned to the **Fn2** button

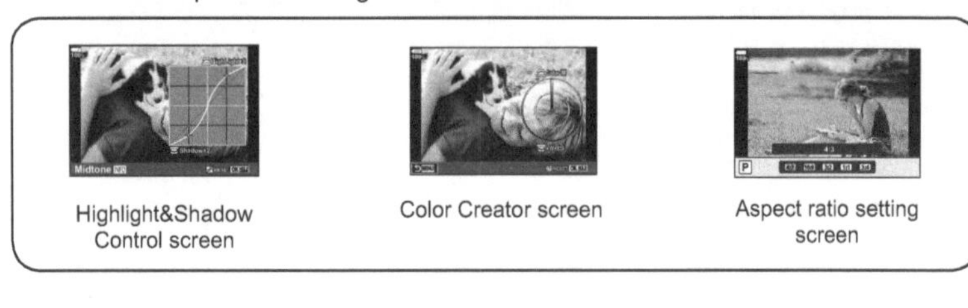

Highlight&Shadow Control screen

Color Creator screen

Aspect ratio setting screen

Function	Front dial	Rear dial
(Highlight&Shadow Control) (P. 66)	Highlight control	Shadow control
(Color Creator) (P. 71)	Hue	Saturation
(ISO/WB) (P. 42, 51/P. 42, 52)	ISO sensitivity	WB mode
(WB/ISO)* (P. 42, 52/P. 42, 51)	WB mode	ISO sensitivity
(Magnify) (P. 41)	Zoom AF: Zoom in or out	
(Image Aspect) (P. 54)	Aspect ratio	
(S-OVF) (P. 121)	—	
(Peaking) (P. 124)	—	

* The function is displayed when it is selected in [Multi Function Settings] (P. 115).

Adjusting overall color (Color Creator)

The overall color of the image can be adjusted using any combination of 30 hues and 8 saturation levels.
You must first set Color Creator to the **Fn2** button (P. 70).

1 Press the **Fn2** button.

- The setting screen will be displayed.

2 Set hue using the front dial and saturation using the rear dial.

- Settings can be reset by pressing and holding the OK button.
- To exit without setting the Color Creator, press the **MENU** button.

3 Press the OK button.

- To return to the Color Creator setting screen, press the **Fn2** button.

- Photos are recorded in RAW+JPEG format when [RAW] is selected for image quality (P. 55, 88).
- Pictures taken using [HDR] (P. 49, 95) or [Multiple Exposure] (P. 95) are recorded at the [Natural] setting.

Shooting “My Clips”

You can create a single My Clips movie file including multiple short movies (clips). You can also add still pictures to the My Clips movie.

■ Shooting

2 Shooting

1 Set the mode dial to 🎥.

2 Press the ⓞⓚ button to display the LV super control panel.

3 Use △▽◁▷ to select [🎥◀︎▪︎].

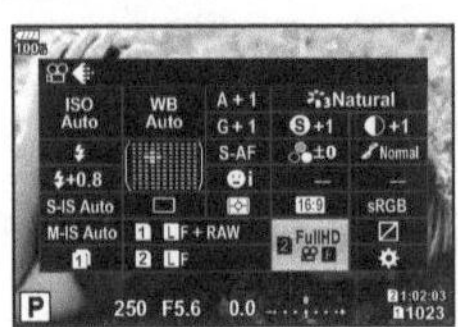

4 Select FHD F 30p using the front dial and press the ⓞⓚ button.

- Press the **INFO** button to change [Movie Resolution], [Bit Rate], [Frame Rate], and [Clip Recording Time]. Highlight [Movie Resolution], [Bit Rate], [Frame Rate], or [Clip Recording Time] using ◁▷ and use △▽ to change.

5 Press the ◉ button to start shooting.

- The recording ends automatically when the preset shooting time has elapsed, and a screen for checking the recorded movie is displayed. Press the ⓞⓚ button to start shooting the next one.
- If you press the ◉ button again during shooting, the shooting will continue while the button is held down (up to a maximum of 16 seconds).

6 Press the ◉ button to shoot the next clip.

- The confirmation screen will disappear and shooting of the next clip will start.
- To delete the clip you have shot or save it in a different My Clips, perform the following on the confirmation screen.

△	Plays back My Clips from the beginning.
▽	Changes My Clips to save the clip to and the position to add the clip to. Use ◁▷ to change the position to add the clip to in the My Clips.
🗑	Deletes the shot clip.

- The shooting of the next clip can be started by pressing the shutter button halfway. The clip is saved in the same My Clips as the previous clip.
- Clips with different [Movie Resolution], bit rate, and [Frame Rate] settings are saved as separate My Clips.

Creating new My Clips

Move a clip to 🎬 using △▽, and press the ⓞ button.

Removing a clip from My Clips

Move a clip to 🎥 using △▽◁▷, and press the ⓞ button.

- The clip removed from My Clips will be a normal movie file.

■ Playback

You can play back the files in My Clips consecutively.

1 Press the ▶ button and select an image marked 🎬.

2 Press the ⓞ button and select [Play My Clips] using △▽. Then press the ⓞ button again.

- The files in My Clips will play back consecutively.
- Press the ⓞ button to end the consecutive playback.

Editing "My Clips"

You can create a single movie file from My Clips.
The shot clips are stored in My Clips. You can add movie clips and still pictures to My Clips. You can also add screen transition effects and art filter effects.

1 Press the ▶ button and then rotate the rear dial to play back My Clips.

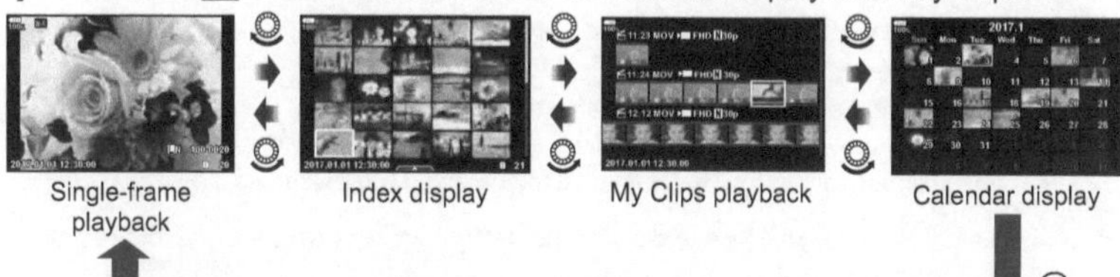

- Images marked with 🎬 in the My Clips playback display can be viewed by pressing the ▶ button, highlighting the image, and pressing ⓞ button.

2 Use △▽ to select My Clips and ◁▷ to select a clip, and press the ⓞ button.

3 Select an item using △▽ and press the ⓚ button.

Play My Clips	Plays back the files in My Clips in order, starting from the beginning.
Movie Interval	Plays back the selected clips as movie.
Rearrange Order	Moves or adds files in My Clips.
Preset Destination	From the next time you shoot, movies shot in the same settings will be added to this My Clips.
Delete My Clips	Deletes all unprotected files from My Clips.
Erase	Select [Yes] and press the ⓚ button to delete.

4 Display My Clips from which you wish to create the movie. Select [Export My Clips] and press the ⓚ button.

5 Select an item using △▽ and press the ⓚ button.

Clip Effects	You can apply 6 types of art effects.
Transition Effect	You can apply fade effects.
BGM	You can set [Party Time] or [Off].
Recorded Clip Volume	When [BGM] is set to [Party Time], you can set a volume for sounds recorded in the movie.
Recorded Clip Sound	By setting to [On], you can create a movie with the recorded sound. This setting is only available when [BGM] is set to [Off].
Preview	You can preview the files of the edited My Clips in order, starting from the first file.

6 When you have finished the editing, select [Begin Export] and press the ⓚ button.

- The combined album is saved as a single movie.
- Exporting a movie may take some time.
- The maximum length for My Clips is 15 minutes and the maximum file size is 4 GB.

- It may take a while for My Clips to display after taking out, inserting, erasing, or protecting the card.
- You can record a maximum of 99 My Clips and a maximum of 99 cuts per clip. The maximum values may vary depending on the file size and length of My Clips.
- You cannot add movies other than clips to My Clips.

- You can change [Party Time] to different BGM. Record the data downloaded from the Olympus website onto the card, select [Party Time] from [BGM] in step 5, and press ▷. Visit the following website for the download.
  http://support.olympus-imaging.com/bgmdownload/

Shooting slow/quick motion movies

You can create slow motion or quick motion movies. You can set the recording speed using 🎥c in record mode.

1 Press the (OK) button to display the LV super control panel.

2 Use △▽◁▷ to select [🎥◀︎].

3 Use the front dial to select C4K 🎥C 24p (custom record mode) (P. 56), then press the (OK) button.

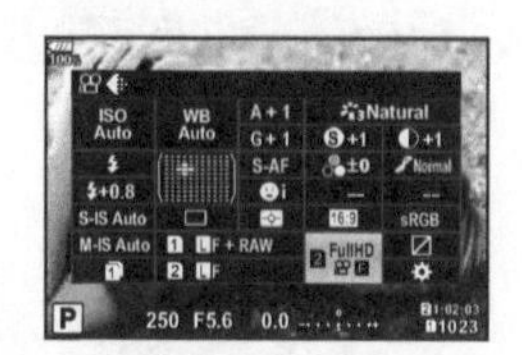

- The option selected for [Slow or Fast Motion] can be changed by pressing the **INFO** button. In ◁▷, select [Slow or Fast Motion] and then press △▽ to select a magnification factor and press the (OK) button. Increase the multiplication factor to shoot a quick motion movie. Reduce the multiplication factor to shoot a slow motion movie. The frame rate changes accordingly.

4 Press the ◉ button to start shooting.

- Press the ◉ button again to end shooting.
- The movie will be played back at a fixed speed so that it appears to be in slow motion or quick motion.

- Sound will not be recorded.
- Any picture mode art filters will be canceled.
- Either or both of slow motion and quick motion cannot be set for some options of [🎥◀︎].
- The shutter speed is limited to values faster than 1/24 seconds when using autofocus. This limitation will not be applied when using manual focus.
- The shutter speed is limited to values faster than 1/24 seconds when using **P**, **A**, or **S** mode in movie shooting.
- When using **M** mode in movie shooting, the shutter speed is limited according to autofocus or manual focus. The focus mode cannot be switched from manual to autofocus.

3 Playback

Information display during playback

Playback image information

Simplified display

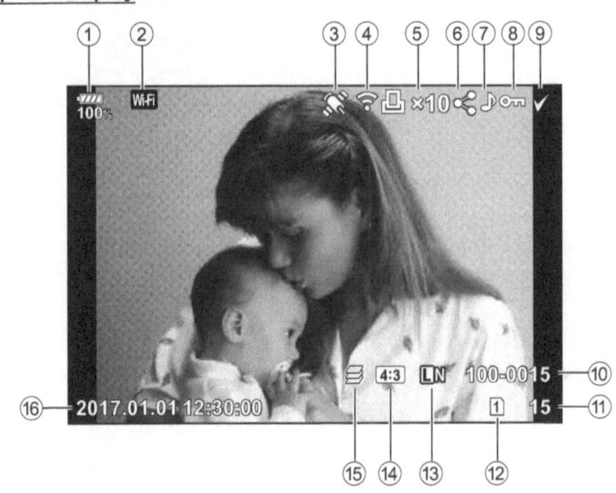

Overall display

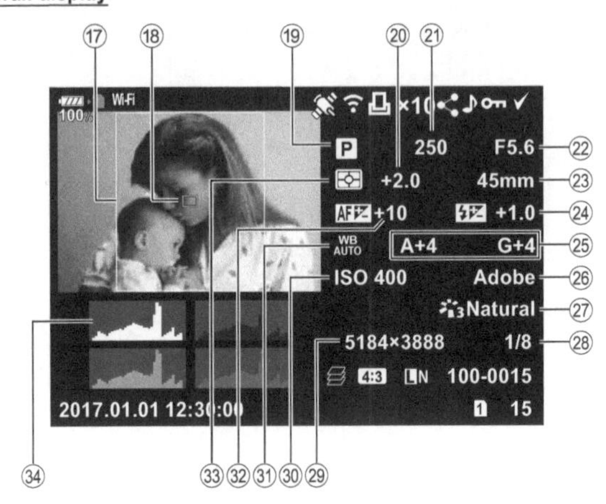

① Battery level....................................P. 18
② Wireless LAN connection......P. 134–138
③ Including GPS information............P. 137
④ Eye-Fi upload complete................P. 122
⑤ Print order
Number of prints...........................P. 144
⑥ Share order.....................................P. 82
⑦ Sound record..........................P. 83, 107
⑧ Protect..P. 81
⑨ Image selected ....P. 82
⑩ File number...................................P. 120
⑪ Frame number
⑫ Slot selected for playback...............P. 78
⑬ Image quality ....P. 55, 88
⑭ Aspect ratio....................................P. 54
⑮ Focus Stacking/
HDR1 HDR2 HDR image.................P. 49, 95
⑯ Date and time ....P. 19
⑰ Aspect border ....P. 54
⑱ AF area pointer...............................P. 40
⑲ Shooting mode......................... P. 24–37
⑳ Exposure compensation ....P. 39
㉑ Shutter speed .... P. 26–29
㉒ Aperture value .... P. 26–29
㉓ Focal length
㉔ Flash intensity control.....................P. 60
㉕ White balance compensation..........P. 52
㉖ Color space.....................................P. 65
㉗ Picture mode.............................P. 61, 88
㉘ Compression rate ....P. 131
㉙ Pixel count ....P. 131
㉚ ISO sensitivity...........................P. 42, 51
㉛ White balance...........................P. 42, 52
㉜ Focus adjustment ....P. 112
㉝ Metering mode...........................P. 45, 51
㉞ Histogram ....P. 23

Switching the information display

You can switch the information displayed during playback by pressing the **INFO** button.

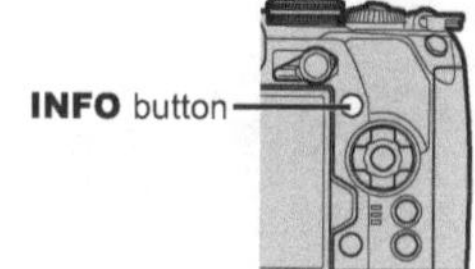

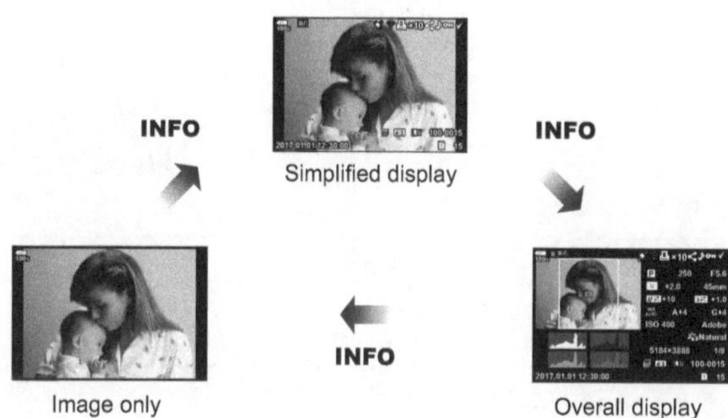

Simplified display

Image only

Overall display

- You can add histogram, highlight & shadow, and Light Box to the information displayed during playback. ☞ [▶ Info] (P. 127)

Viewing photographs and movies

1 Press the ▶ button.

- Your most recent photograph or movie will be displayed.
- Select the desired photograph or movie using the front dial (front dial icon) or arrow pad.
- Press the shutter button halfway to return to shooting mode.

Tips

- To change the playback card, while pressing the ▶ button, turn the dial to select a slot, then release the ▶ button. The playback card setting will return to the original setting once the camera exits the playback mode. The change of playback card by this operation does not affect the [▶ Slot] setting of [Card Slot Settings] (P. 132).

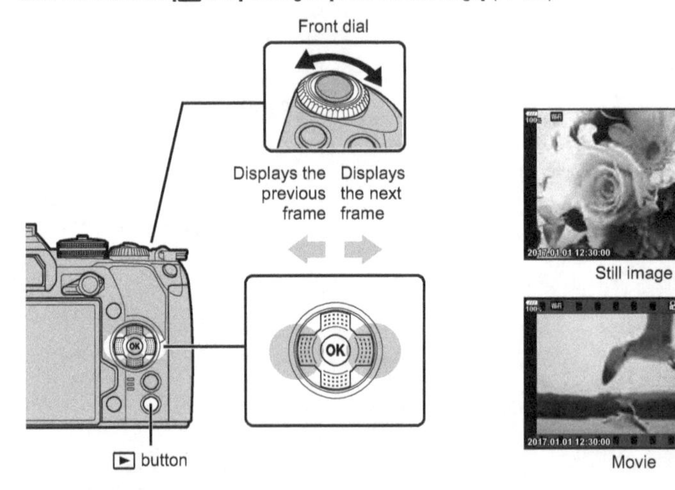

Still image
Movie

Rear dial (rear dial icon)	Zoom in (rear dial icon)/Index (rear dial icon)
Front dial (front dial icon)	Previous (front dial icon)/Next (front dial icon) Operation also available during close-up playback.
Arrow pad (△▽◁▷)	Single-frame playback: Next (▷)/previous (◁)/playback volume (△▽) Close-up playback: Changing the close-up position You can display the next frame (▷) or the previous frame (◁) during close-up playback by pressing the **INFO** button. Press the **INFO** button again to display a zoom frame and use △▽◁▷ to change its position. Index/My Clips/calendar playback: Highlight image
INFO	View image info
☑	Select picture (P. 82)
AEL/AFL button	Protect picture (P. 81)

🗑	Delete picture (P. 82)
ⓞ	View menus (in calendar playback, press this button to exit to single-frame playback)

Index display/Calendar display

- In single-frame playback, rotate the rear dial to ▦ for index playback. Rotate further for My Clips playback and further still for calendar playback.
- Turn the rear dial to Q to return to single-frame playback.

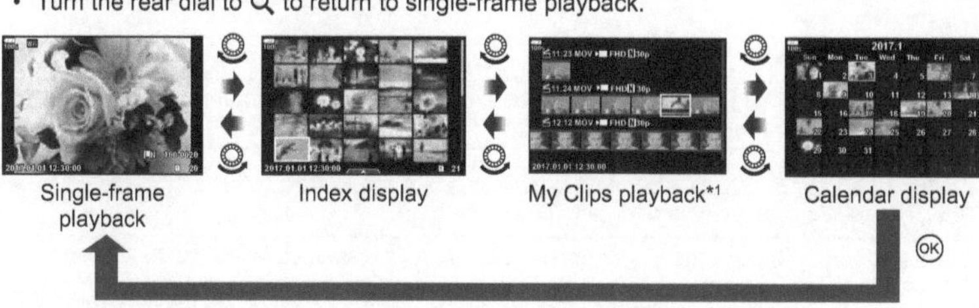

Single-frame playback | Index display | My Clips playback*1 | Calendar display

*1 If one or more My Clips have been created, it will be displayed here (P. 72).

- You can change the number of frames for index display. ☞ [▦ Settings] (P. 128)

Viewing still images

Close-up playback

In single-frame playback, turn the rear dial to Q to zoom in. Turn to ▦ to return to single-frame playback.

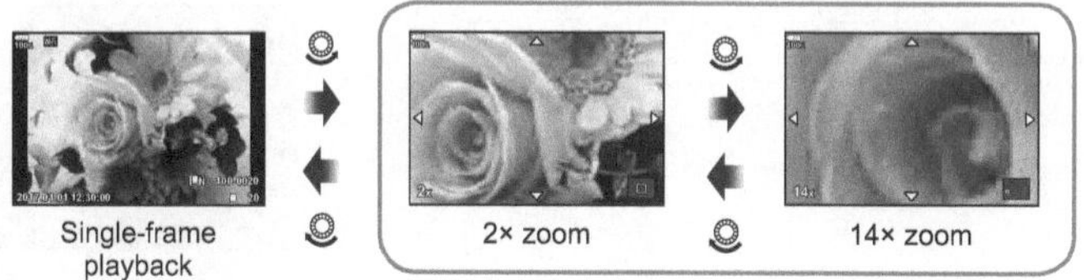

Single-frame playback | 2× zoom | 14× zoom

Close-up playback

Rotate

Choose whether to rotate photographs.

1 Play the photograph back and press the ⓞ button.

2 Select [Rotate] and press the ⓞ button.

3 Press △ to rotate the image counterclockwise, ▽ to rotate it clockwise; the image rotates each time the button is pressed.

- Press the ⓞ button to save settings and exit.
- The rotated image is saved in its current orientation.
- Movies and protected images cannot be rotated.

Slideshow

This function displays images stored on the card one after another.

1 Press the ⓚ button during playback and select [ ].

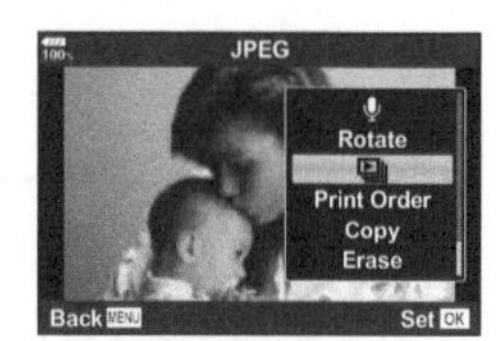

2 Adjust settings.

Start	Start the slideshow. Images are displayed in order, starting with the current picture.
BGM	Set [Party Time] or [Off].
Slide	Set the type of data to be played back.
Slide Interval	Choose the length of time each slide is displayed from 2 to 10 seconds.
Movie Interval	Select [Full] to play back full length of each movie clip in the slideshow or [Short] to play back only the opening portion of each movie clip.

3 Select [Start] and press the ⓚ button.

- The slideshow will start.
- Press the ⓚ button to stop the slideshow.

Volume

Volume can be adjusted by pressing △ or ▽ during single-frame and movie playback.

> **Slideshow volume**
> Press △▽ during the slideshow to adjust the overall volume of the camera speaker. Press ◁▷ while the volume adjustment indicator is displayed to adjust the balance between the sound recorded with the image or movie and background music.

3 Playback

Watching movies

Select a movie and press the ⓞ button to display the playback menu. Select [Play Movie] and press the ⓞ button to begin playback. Fast-forward and rewind using ◁/▷. Press the ⓞ button again to pause playback. While playback is paused, use △ to view the first frame and ▽ to view the last frame. Use ◁▷ or the front dial (🎛) to view previous or next frame. Press the **MENU** button to end playback.

For movies of 4 GB or larger

If the movie was automatically split into multiple files, pressing ⓞ will display a menu containing the following options:

[Play from Beginning]:	Plays back a split movie all the way through
[Play Movie]:	Plays back files separately
[Delete entire 🎥]:	Deletes all parts of a split movie
[Erase]:	Deletes files separately

- We recommend using the latest version of OLYMPUS Viewer 3 to play back movies on a computer. Before launching the software for the first time, connect the camera to the computer.

Protecting images

Protect images from accidental deletion.
Display the image you want to protect and press the **AEL/AFL** button to add O‑n to the image (protection icon). Press the **AEL/AFL** button again to cancel protection. You can also protect multiple selected images. ☞ "Selecting images (O‑n, Copy Select, Erase Selected, Share Order Selected)" (P. 82)

- Formatting the card erases all data including protected images.

Copying an image (Copy)

When there are cards with available space in both slots 1 and 2, you can copy an image to the other card. When playing back an image you want to copy, press the ⓞ button to display the playback menu. After selecting [Copy] and pressing the ⓞ button, select whether or not to specify a folder to save to. Select [Yes], then press the ⓞ button to copy the image to the other card.

- You can also copy all images on a card to the other card at a time. ☞ "Copy All" (P. 108)

Erasing an image

Display an image you want to delete and press the 🗑 button. Select [Yes] and press the Ⓞ button.
You can erase images without the confirmation step by changing the button settings.
☞ [Quick Erase] (P. 121)

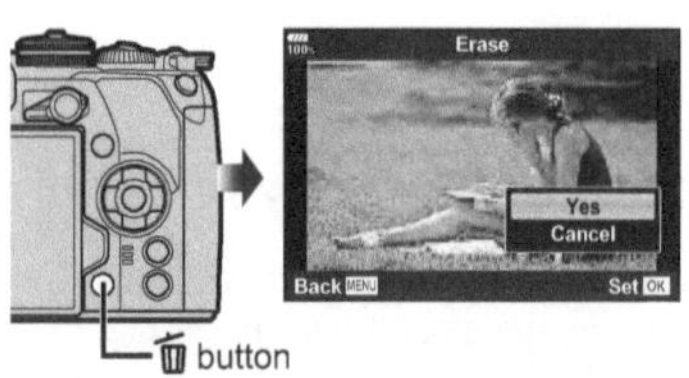

Selecting images (O‑n, Copy Select, Erase Selected, Share Order Selected)

You can select multiple images for [O‑n], [Copy Select], [Erase Selected] or [Share Order Selected].
Press the ☑ button on the index display screen (P. 79) to select an image; a ✔ icon will appear on the image. Press the ☑ button again to cancel the selection.
Press the Ⓞ button to display the menu, and then select from [O‑n], [Copy Select], [Erase Selected] or [Share Order Selected].

Setting a transfer order on images (Share Order)

You can select images you want to transfer to a smartphone in advance. You can also browse the images included in the share order. When playing back images you want to transfer, press the Ⓞ button to display the playback menu. After selecting [Share Order] and pressing the Ⓞ button, press △ or ▽ to set a share order on an image and display <. To cancel a share order, press △ or ▽.
You can select images you want to transfer in advance and set a share order all at once. ☞ "Selecting images (O‑n, Copy Select, Erase Selected, Share Order Selected)" (P. 82), "Transferring images to a smartphone" (P. 136)

- You can set a share order on 200 frames.
- Share orders cannot include RAW images or Motion JPEG (HD) movies.

Audio recording

Audio can be added to still images (up to 30 sec. long).

1 Display the image to which you want to add audio and press the ⓞ button.
- Audio recording is not available with protected images.
- Audio recording is also available in the playback menu.

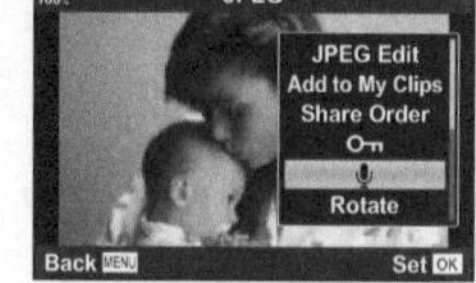

2 Select [🎙] and press the ⓞ button.
- To exit without adding audio, select [Cancel].

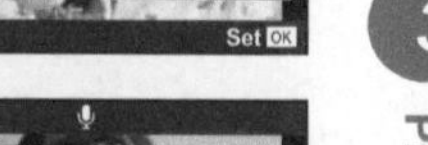

3 Select [🎙 Start] and press the ⓞ button to begin recording.

4 Press the ⓞ button to end recording.
- Images recorded with audio are indicated by a ♪ icon.
- To delete recorded audio, select [Erase] in Step 3.

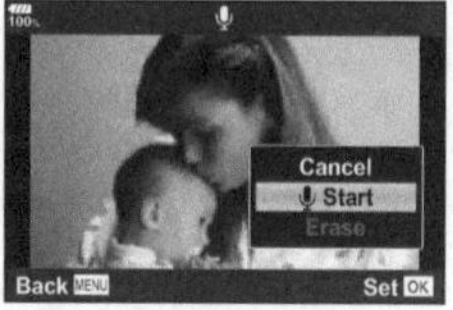

Adding still pictures to My Clips (Add to My Clips)

You can also select still pictures and add them to My Clips.
Display the still picture you wish to add and press the ⓞ button to display the menu. Select [Add to My Clips] and press the ⓞ button. Using △▽◁▷, select My Clips and the order in which you wish to add the pictures, then press the ⓞ button.
- For RAW or High Res Shot images, coarse images used for display are added.

Using the touch screen

You can use the touch screen to manipulate images.

■ Full-frame playback

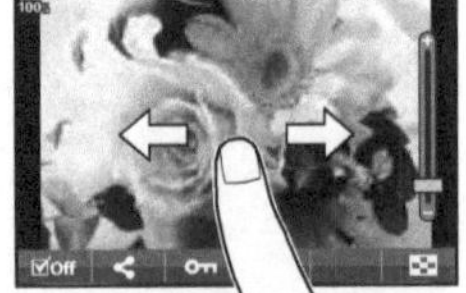

Displaying the previous or next image

- Slide your finger to the left to view the next image, and right to view the previous image.

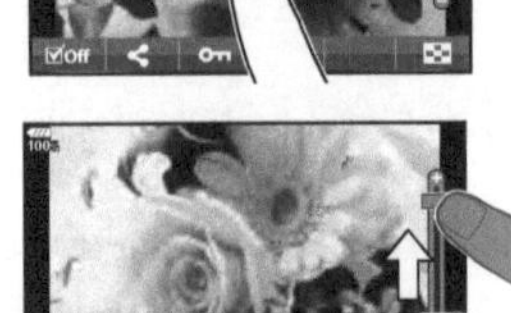

Magnify

- Lightly tap the screen to display the slider and [icon].
- Slide the bar up or down to zoom in or out.
- Slide your finger to scroll the display when the picture is zoomed in.
- Tap [icon] to display index playback.
  Tap [icon] for calendar playback and My Clips playback.

■ Index/My Clips/Calendar playback

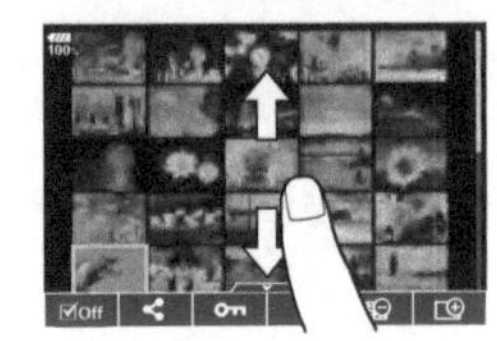

Displaying the previous or next page

- Slide your finger up to view the next page, and down to view the previous page.
- Tap [icon] or [icon] to switch the number of images displayed. ☞ [[icon] Settings] (P. 115)
- Tap [icon] several times to return to single-frame playback.

Viewing images

- Tap an image to view it full frame.

Selecting and protecting images

In single-frame playback, lightly tap the screen to display the touch menu. You can then perform the desired operation by tapping the icons in the touch menu.

☑	Select an image. You can select multiple images and delete them collectively.
<	Images you want to share with a smartphone can be set. ☞ "Setting a transfer order on images (Share Order)" (P. 82)
O‑n	Protects an image.

- Do not tap the display with your fingernails or other sharp objects.
- Gloves or monitor covers may interfere with touch screen operation.

3 Playback

4 Menu functions

Basic menu operations

The menus contain shooting and playback options that are not displayed by the LV super control panel, etc., and let you customize the camera settings for easier use.

1	Preliminary and basic shooting options (P. 86)
2	Advanced shooting options (P. 86)
	Movie mode settings (P. 100)
	Playback and retouch options (P. 105)
	Customizing camera settings (P. 111)
	Camera setup (e.g., date and language) (P. 109)

1 Press the **MENU** button to display the menus.

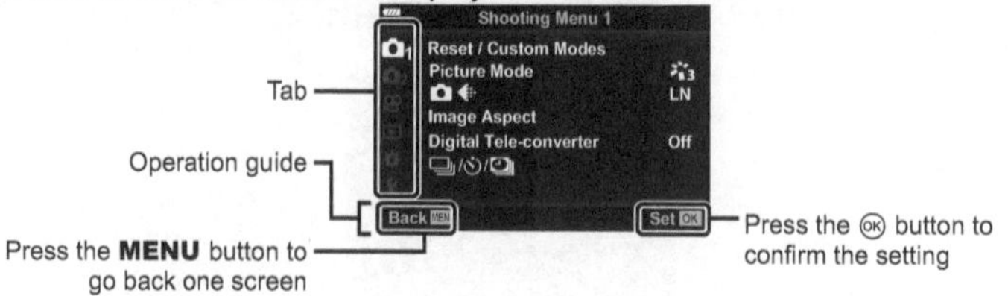

- A guide is displayed in 2 seconds after you select an option.
- Press the **INFO** button to view or hide guides.

2 Use △▽ to select a tab and press the OK button.

- The menu group tab appears when the Custom Menu is selected. Use △▽ to select the menu group and press the OK button.

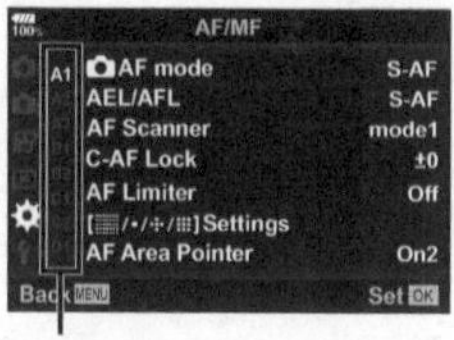

3 Select an item using △▽ and press the ⊛ button to display options for the selected item.

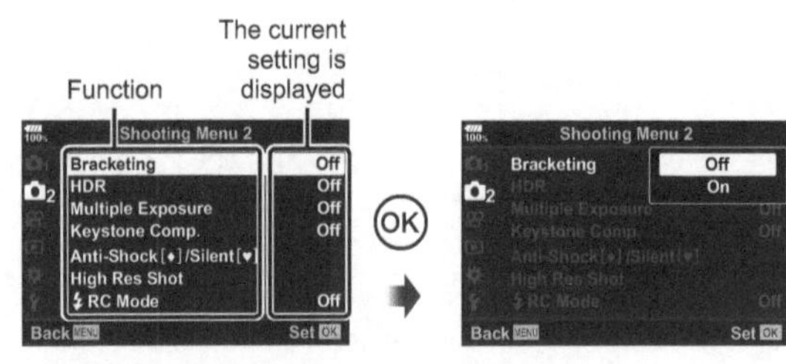

4 Use △▽ to highlight an option and press the ⊛ button to select.

- Press the **MENU** button repeatedly to exit the menu.

- For the default settings of each option, refer to "Menu directory" (P. 165).

Using Shooting Menu 1/Shooting Menu 2

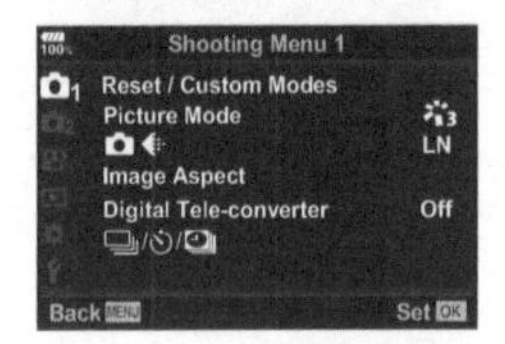

Shooting Menu 1

Reset / Custom Modes (P. 86)
Picture Mode (P. 61, 88)
(P. 55, 88)
Image Aspect (P. 54)
Digital Tele-converter (P. 88)
(Sequential shooting/ Self-timer/Time lapse shooting) (P. 46, 54, 89, 90)

Shooting Menu 2

Bracketing (P. 91)
HDR (P. 49, 95)
Multiple Exposure (P. 95)
Keystone Comp. (P. 97)
Anti-Shock [♦]/Silent [♥] (P. 98)
High Res Shot (P. 99)
RC Mode (P. 99, 153)

Returning to default settings (Reset)

Camera settings can be easily restored to default settings.

1 Select [Reset / Custom Modes] in Shooting Menu 1 and press the ⊛ button.

2 Select [Reset] and press the ⊛ button.

- Highlight [Reset] and press ▷ to choose the reset type. To reset all settings except the time, date, and a few others, highlight [Full] and press the ⊛ button. ☞ "Menu directory" (P. 165)

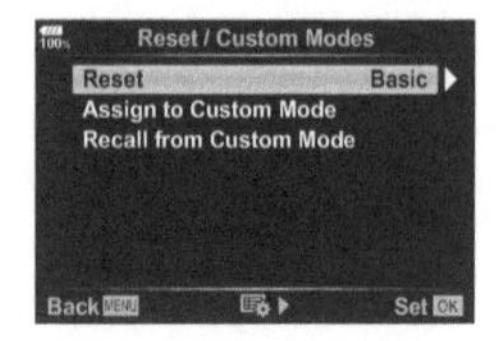

3 Select [Yes] and press the ⊛ button.

Registering favorites settings (Assign to Custom Mode)

Editing custom settings

Current camera settings can be saved to three Custom Modes (C1 to C3). Saved settings can be recalled by rotating the mode dial to **C1, C2**, or **C3**, or using [Recall from Custom Mode].

1. Adjust settings in order to save.
   - Set the mode dial to the positions other than iAUTO (iAUTO), **ART**, or movie (🎥) mode.
2. Select [Reset/Custom Modes] in 📷 Shooting Menu 1 and press the ⓞⓚ button.
3. Select [Assign to Custom Mode] and press ▷.
4. Select the desired destination ([Custom Mode C1]–[Custom Mode C3]) and press the ⓞⓚ button.
5. Select [Set] and press the ⓞⓚ button.
   - Selecting [Set] saves current settings, overwriting the registered settings.
   - To cancel the registration, select [Reset].

- Settings that can be saved to the Custom Modes ☞ "Menu directory" (P. 165)
- Settings are preset in each Custom Mode. Selecting [Reset] > [Full] when resetting camera settings restores the preset settings for each Custom Mode. See "Default Custom Mode options" for default settings (P. 175).

Recalling custom settings

Settings saved to mode dial **C1, C2** or **C3** can be recalled.

1. Select [Reset / Custom Modes] in 📷 Shooting Menu 1 and press the ⓞⓚ button.
2. Select [Recall from Custom Mode] and press ▷.
3. Select [Custom Mode C1]–[Custom Mode C3] and press the ⓞⓚ button.
4. Select [Yes] and press the ⓞⓚ button.

- The shooting mode will not be applied if the mode dial is set to **P**, **A**, **S** or **M**.

Processing options (Picture Mode)

You can make individual adjustments to contrast, sharpness and other parameters in [Picture Mode] (P. 61) settings. Changes to parameters are stored for each picture mode separately.

1 Select [Picture Mode] in Shooting Menu 1 and press the OK button.

- The camera will display the picture mode available in the current shooting mode.

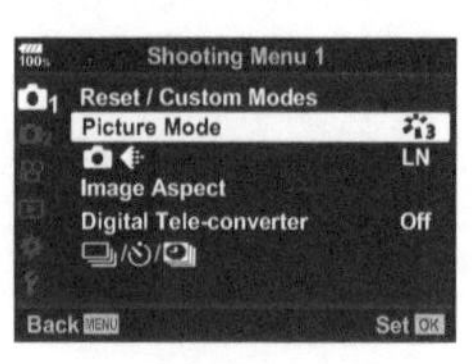

2 Select an option with △▽ and press the OK button.

- Press ▷ to set the detailed options for the selected picture mode. Detailed options are not available for some picture modes.
- Changes to contrast have no effect at settings other than [Normal].

Tips

- You can reduce the number of picture mode options displayed in the menu.
  ☞ [Picture Mode Settings] (P. 115)

Image quality ( )

☞ "Selecting image quality ( )" (P. 55)

- You can change the JPEG image size and compression ratio combination, and [M] and [S] pixel counts. [Set], [Pixel Count] ☞ "Combinations of JPEG image sizes and compression rates" (P. 131)

Digital Zoom (Digital Tele-converter)

Digital Tele-converter is used to zoom in beyond the current zoom ratio. The camera saves the center crop. The subject is nearly doubled in size.

1 Select [On] for [Digital Tele-converter] in Shooting Menu 1.

2 The view in the monitor will be enlarged by a factor of two.

- The subject will be recorded as it appears in the monitor.

- This function cannot be used in multiple exposure shooting.
- This function is not available when the information of [Movie Effect] is displayed on the screen in movie mode.
- This function is not available when [Movie] is assigned to a button with [Button Function].
- When a RAW image is displayed, the area visible in the monitor is indicated by a frame.
- AF target drops.
- is displayed on the monitor.

Setting the custom self-timer (□ᵢ/⏲)

You can customize the self-timer operation.

1 Select [□ᵢ/⏲/▣ᵢ] in 📷₁ Shooting Menu 1 and press the ⓞ button.

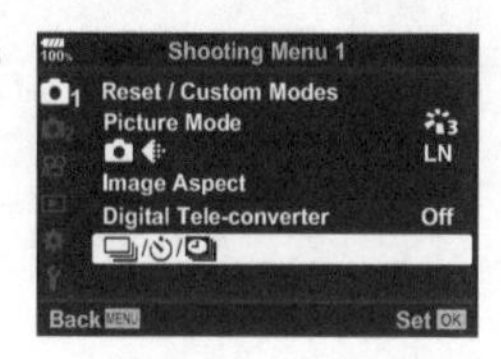

2 Select [□ᵢ/⏲] and press ▷.

3 Select [⏲C] (custom) and press ▷.

4 Use △▽ to select the item and press ▷.

- Use △▽ to select the setting and press the ⓞ button.

Number of Frames	Sets the number of frames to be shot.
⏲ Timer	Sets the time after the shutter button is pressed until the picture is taken.
Interval Length	Sets the shooting interval for the second and subsequent frames.
Every Frame AF	Sets whether or not to perform AF right before a picture is taken with self-timer.

Shooting automatically with a fixed interval (time lapse shooting)

You can set the camera to shoot automatically with a set time lapse. The shot frames can also be recorded as into a single movie. This setting is only available in **P**/**A**/**S**/**M** modes.

1 Select [/ /] in Shooting Menu 1 and press the OK button.

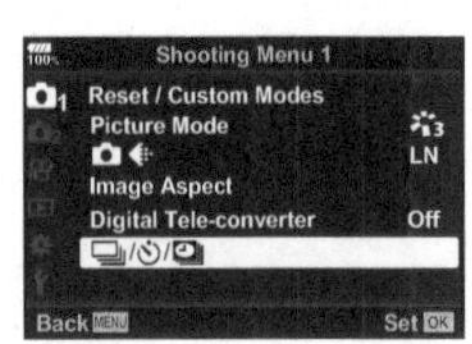

2 Select [Intrvl. Sh./Time Lapse] and press ▷.

3 Select [On] and press the ▷.

4 Adjust the following settings and press the OK button.

Number of Frames	Sets the number of frames to be shot.
Start Waiting Time	Sets waiting time before starting to shoot.
Interval Length	Sets interval between shots after shooting starts.
Time Lapse Movie	Sets recording format of frame sequence. [Off]: Records each frame as a still picture. [On]: Records each frame as a still picture and generates a single movie from the frame sequence.
Movie Settings	[Movie Resolution]: Choose a size for time lapse movies. [Frame Rate]: Choose a frame rate for time lapse movies.

5 Shoot.

- Frames are shot even if the image is not in focus after AF. If you wish to fix the focus position, shoot in MF.
- [Rec View] (P. 109) operates for 0.5 seconds.
- If either of the time before shooting, or shooting interval is set to 1 minute 31 seconds or longer, the monitor and camera power will turn off after 1 minute. 10 seconds before shooting, the power will automatically turn on again. When the monitor is off, press the shutter button to turn it on again.
- If the AF mode (P. 43, 51) is set to [C-AF] or [C-AF+TR], it is automatically changed to [S-AF].
- Touch operations are disabled during time lapse shooting.
- This function cannot be used with HDR photography.
- It is not possible to combine time lapse shooting with bracketing, multiple exposure, and bulb, time or composite photography.
- The flash will not work if the flash charging time is longer than the interval between shots.
- If the camera automatically turns off in interval between shots, it will turn on in time for the next shot.
- If any of still pictures are not recorded correctly, the time lapse movie will not be generated.
- If there is insufficient space on the card, the time lapse movie will not be recorded.
- Time lapse shooting will be canceled if any of the following is operated:
  Mode dial, **MENU** button, ▶ button, lens release button, or connecting the USB cable.
- If you turn off the camera, time lapse shooting will be canceled.

- If there is not enough charge left on the battery, the shooting may end partway through. Make sure the battery is charged enough before shooting.
- Depending on your computer's system environment, you may not be able to view [4K] movies on your computer. More information is available on the OLYMPUS website.

Varying settings over a series of photographs (Bracketing)

"Bracketing" refers to the act of varying settings automatically over a series of shots or a series of images to "bracket" the current value. You can store bracket shooting settings and turn bracket shooting off.

1 Select [Bracketing] in Shooting Menu 2 and press the OK button.

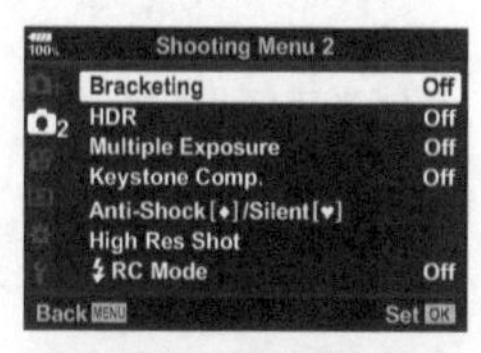

2 After selecting [On], press ▷ and select a bracket shooting type.

- When you select bracket shooting, BKT appears on the screen.

3 Press ▷, select settings for parameters such as the number of shots, and then press the OK button.

- Press the OK button repeatedly until the screen returns to the one in step 1.
- If you select [Off] in step 2, bracket shooting settings will be saved and you can shoot normally.

- Bracketing can not be combined with HDR, interval-timer photography, digital shift, multiple-exposure photography, or high res shots.
- Bracketing is not available if there is not enough space on the camera's memory card for the selected number of frames.

Tips

- If you turn the Custom Menu [Switch Function] (P. 113) [On], bracketing settings are available with button operation. Set the **Fn** lever to position 2 and rotate the dial while pressing the **HDR** button. You can select the bracket shooting type with the front dial and the number of shots with the rear dial. After settings are complete, you can switch between bracketing and normal shooting by pressing the **HDR** button.

AE BKT (AE bracketing)

The camera varies the exposure of each shot. You can select the bracketing increment from 0.3 EV, 0.7 EV, and 1.0 EV. In single-frame shooting mode, one photograph is taken each time the shutter button is pressed all the way down, while in sequential shooting mode the camera continues to take shots in the following order while the shutter button is pressed all the way down: no modification, negative, positive. Number of shots: 2, 3, 5, or 7

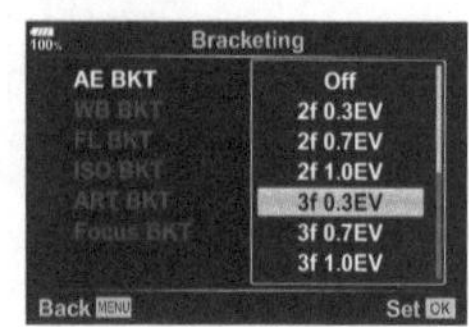

- The BKT indicator turns green during bracketing.
- The camera modifies exposure by varying aperture and shutter speed (**P** mode), shutter speed (**A** and **M** modes), or aperture (**S** mode). If [All] is selected for [ISO-Auto] (P. 117) in **M** mode and [AUTO] is selected for [ISO] (P. 42, 51), the camera modifies exposure by varying ISO sensitivity.
- The camera brackets the value currently selected for exposure compensation.
- The size of the bracketing increment changes with the value selected for [EV Step]. ☞ [EV Step] (P. 117)

WB BKT (WB bracketing)

Three images with different white balances (adjusted in specified color directions) are automatically created from one shot, starting with the value currently selected for white balance. WB bracketing is available in **P**, **A**, **S** and **M** modes.

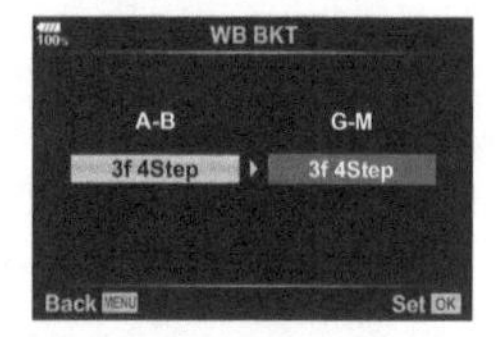

- White balance can be varied by 2, 4, or 6 steps on each of the A–B (Red–Blue) and G–M (Green–Magenta) axes.
- The camera brackets the value currently selected for white balance compensation.

FL BKT (FL bracketing)

The camera varies flash level over three shots (no modification on the first shot, negative on the second, and positive on the third). In single-frame shooting, one shot is taken each time the shutter button is pressed; in sequential shooting, all shots are taken while the shutter button is pressed.

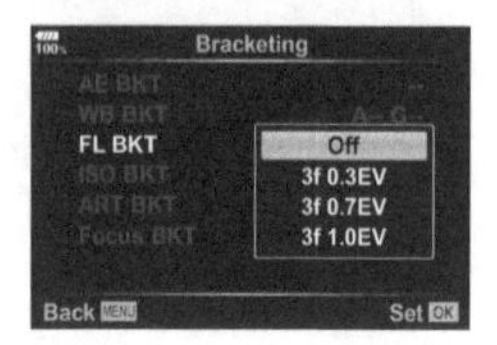

- The BKT indicator turns green during bracketing.
- The size of the bracketing increment changes with the value selected for [EV Step]. ☞ [EV Step] (P. 117)

ISO BKT (ISO bracketing)

The camera varies the sensitivity over three shots while keeping the shutter speed and aperture fixed. You can select the bracketing increment from 0.3 EV, 0.7 EV, and 1.0 EV. Each time the shutter button is pressed, the camera shoots three frames with the set sensitivity (or if auto sensitivity is selected, the optimal sensitivity setting) on the first shot, negative modification on the second shot, and positive modification on the third shot.

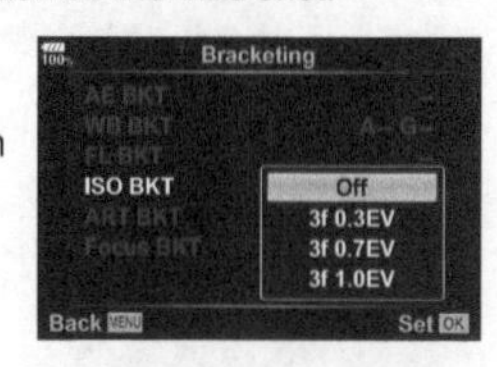

- The size of the bracketing increment does not change with the value selected for [ISO Step]. ☞ [ISO Step] (P. 117)
- Bracketing is performed regardless of the upper limit set with [ISO-Auto Set]. ☞ [ISO-Auto Set] (P. 117)

ART BKT (ART bracketing)

Each time the shutter is released, the camera records multiple images, each with a different art filter setting. You can turn art filter bracketing on or off separately for each picture mode.

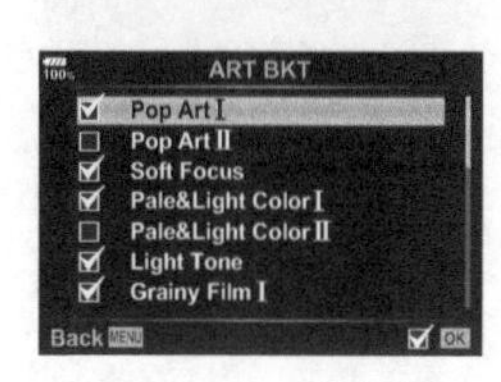

- Recording may take some time.
- ART BKT cannot be combined with WB BKT or ISO BKT.

Focus BKT (Focus bracketing)

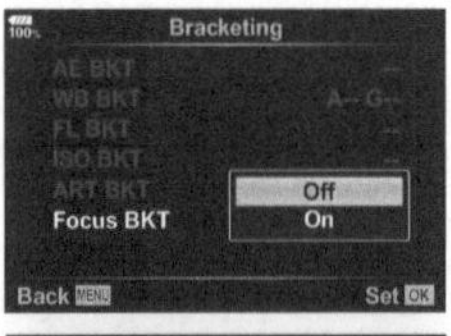

Take a series of shots at different focus positions. Focus moves successively farther from the initial focus position. Choose the number of shots using [Set number of shots] and the change in focus distance using [Set focus differential]. Choose smaller values for [Set focus differential] to narrow the change in focus distance, larger values to widen it. If you are using a flash other than the dedicated flash unit, you can specify the time it takes to charge using the [⚡Charge Time] option.

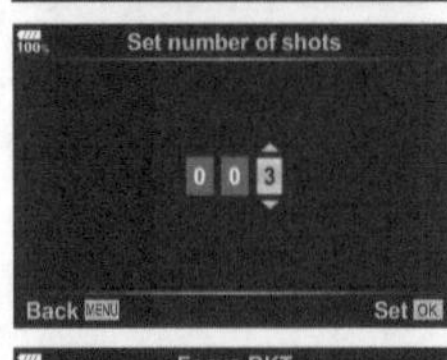

Press the shutter button all the way down and release it immediately. Shooting will continue until the selected number of shots is taken or until the shutter button is pressed all the way down again.

- Focus bracketing is not available with lenses that have mounts conforming to the Four-Thirds standard.
- Focus bracketing ends if zoom or focus is adjusted during shooting.
- Shooting ends when focus reaches infinity.
- Pictures taken using focus bracketing are shot in silent mode.
- To use the flash, select [Allow] for [Silent [♥] Mode Settings] > [Flash Mode].
  ☞ [Silent [♥] Mode Settings] (P. 98)
- Focus bracketing can not be combined with other forms of bracketing.

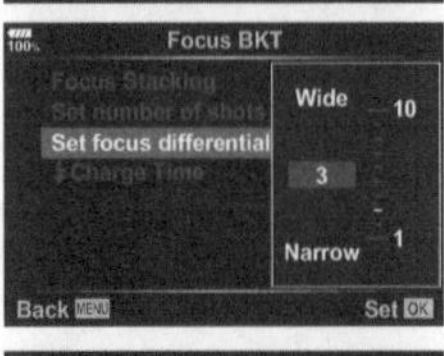

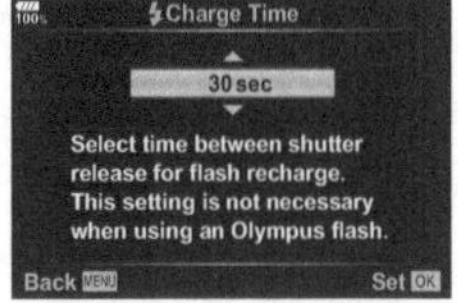

Focus BKT (Focus Stacking)

The focus position is automatically shifted to capture 8 shots which are then composited for a single JPEG image that is in focus all the way from the foreground to background.

- The focus position is automatically shifted based on the center of the focal position and 8 frames are captured in a single shot.
- If compositing fails, the image will not be saved.
- Focus stacking ends if zoom or focus is adjusted during shooting.
- The angle of view for composited images is narrower than the original images.
- See the OLYMPUS website for information on the lenses that can be used with [Focus Stacking].
- Focus stacking can not be combined with other forms of bracketing.

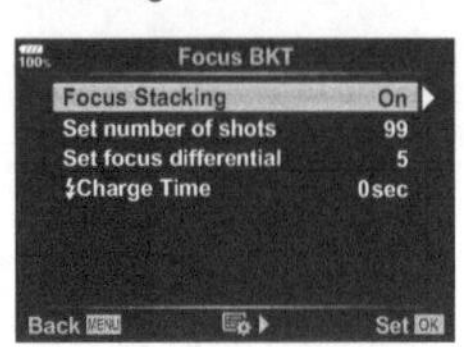

Taking HDR (High Dynamic Range) images

You can shoot **HDR** (High Dynamic Range) images.
☞ "Taking HDR (High Dynamic Range) images" (P. 49)

1 Select [HDR] in 🗗2 Shooting Menu 2 and press the ⓞ button.

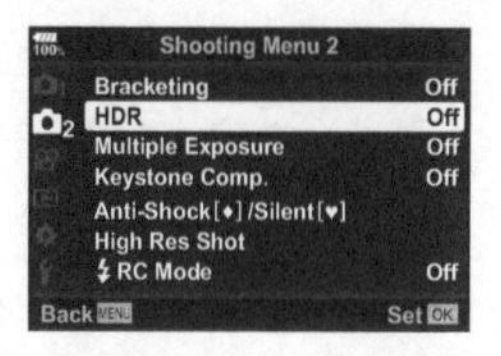

2 Select a type of HDR photography and press the ⓞ button.

3 Shoot.

- When you press the shutter button, the camera automatically shoots the set number of images.

Recording multiple exposures in a single image (Multiple Exposure)

Record multiple exposures in a single image, using the option currently selected for image quality.

1 Select [Multiple Exposure] in 🗗2 Shooting Menu 2 and press the ⓞ button.

2 Use △▽ to select the item and press ▷.

- Use △▽ to select the setting and press the ⓞ button.

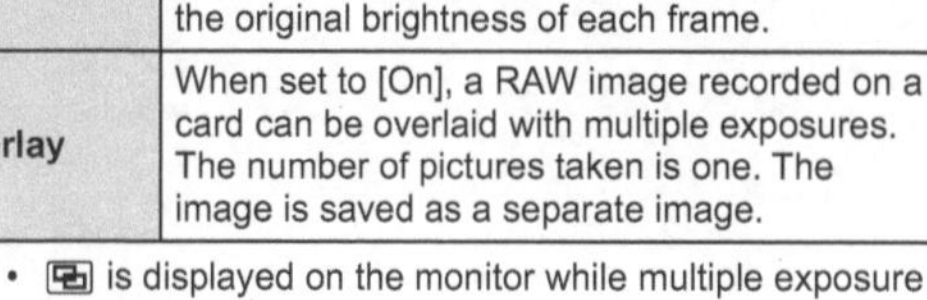

Number of Frames	Select [2f].
Auto Gain	When set to [On], the brightness of each frame is set to 1/2 and the images are overlaid. When set to [Off], the images are overlaid with the original brightness of each frame.
Overlay	When set to [On], a RAW image recorded on a card can be overlaid with multiple exposures. The number of pictures taken is one. The image is saved as a separate image.

- ▣ is displayed on the monitor while multiple exposure is in effect.

3 Shoot.

- ▣ is displayed in green if shooting starts.
- Press 🗑 to delete the last shot.
- The previous shot is superimposed on the view as a guide to framing the next shot.

- The camera will not go to sleep while multiple exposure is in effect.
- Photographs taken with other cameras cannot be included in a multiple exposure.
- When [Overlay] is set to [On], the images displayed when a RAW image is selected are developed with the settings at the time of shooting.
- To set the shooting functions, cancel multiple exposure shooting first. Some functions cannot be set.
- Multiple exposure is canceled automatically from the first picture in the following situations.

  The camera is turned off, the ▶ button is pressed, the **MENU** button is pressed, the shooting mode is set to a mode other than **P**, **A**, **S**, **M**, the battery power runs out, or any cable is connected to the camera.
- When a RAW image is selected using [Overlay], the JPEG image for the image recorded in RAW+JPEG is displayed.
- Multiple exposure cannot be used with some shooting functions such as bracketing.

Tips

- To overlay 3 or more frames: Select RAW for [◀:·] and use the [Overlay] option to make repeated multiple exposures.
- For more information on overlaying RAW images: ☞ "Image overlay" (P. 107)

Keystone correction and perspective control (Keystone Comp.)

Use keystone correction for shots taken from the bottom of a tall building, or deliberately exaggerate the effects of perspective. This setting is only available in **P/A/S/M** modes.

1 Select [On] for [Keystone Comp.] in 2 Shooting Menu 2.

2 Adjust the effect in the display and frame the shot.

- Use the front dial and rear dial for keystone correction.
- Use △▽◁▷ to choose the area to be recorded. The area cannot be changed depending on the amount of correction.
- Press and hold the ⓞ button to cancel any changes.
- To adjust aperture, shutter speed, and other shooting options while keystone compensation is in effect, press the **INFO** button to view a display other than keystone compensation adjustment. To resume keystone compensation, press the **INFO** button until keystone compensation adjustment is displayed.
- The following may occur as the correction amount increases.
  - The image will be coarse.
  - The magnification ratio for image cropping will be large.
  - The crop position will not be able to move.

3 Shoot.

- To end keystone compensation, select [Off] for [Keystone Comp.] in 2 Shooting Menu 2.

- When [(Keystone Comp.)] (P. 68) is assigned to a button using Button Function, press and hold the selected button to end keystone correction.
- Photos are recorded in RAW+JPEG format when [RAW] is selected for image quality.
- The desired results may not be obtained with converter lenses.
- Depending on the amount of correction, some AF targets may be outside the display area. An icon (, , or) is displayed when the camera focuses on an AF target outside the display area.
- The following is not available while keystone compensation is in effect.
  Live bulb/live time/composite photography, sequential shooting, bracketing, HDR, multiple exposure, digital tele-converter, movie, [C-AF] and [C-AF+TR] autofocus modes, [e-Portrait] and **ART** picture modes, custom self-timer, High Res Shot
- If a focus distance is selected for [Image Stabilizer] or you are using a lens for which lens info has been provided, the correction will be adjusted accordingly. Except when using a Micro Four Thirds or Four Thirds lens, choose a focal length using the [Image Stabilizer] option (P. 53).

Setting anti-shock/silent shooting (Anti-Shock [♦]/Silent [♥])

By setting anti-shock/silent shooting you can select anti-shock or silent shooting when using sequential shooting/self-timer (P. 46).

1. Select [Anti-Shock [♦]/Silent [♥]] in Shooting Menu 2 and press the OK button.
2. Use △▽ to select the item and press ▷.
   - Use △▽ to select the setting and press the OK button.

Anti-Shock [♦]	Sets the time period between the shutter button being pressed all the way down and the shutter release when shooting in anti-shock mode. If the interval is set, the item marked with [♦] is displayed as an option for sequential shooting/self-timer mode. When not using anti-shock shooting, set to [Off]. Use this mode to suppress small vibrations caused by the operation of the shutter. Anti-shock mode is available in both sequential shooting and self-timer modes (P. 46).
Silent [♥]	Sets the time period between the shutter button being pressed all the way down and the shutter release when shooting in silent mode. If the inteval is set, the item marked with ♥ is displayed as an option for sequential shooting/self-timer mode. When not using silent shooting, set to [Off].
Noise Reduction [♥]	Set to [Auto] to reduce noise in long exposure shots when using silent shooting mode. During noise reduction processing, the sound of shutter operation can be heard.
Silent [♥] Mode Settings	Choose [Allow] or [Not Allow] for each of [■)))], [AF Illuminator], and [Flash Mode].

Setting high resolution shooting (High Res Shot)

By specifying the setting for [High Res Shot], high resolution shooting can be activated by selecting ▣ from options for sequential shooting/self-timer mode (P. 46).

1 Select [High Res Shot] in ◘₂ Shooting Menu 2 and press the Ⓞ button.

2 Use △▽ to select the item and press ▷.

- Use △▽ to select the setting and press the Ⓞ button.

High Res Shot	Sets the time period between the shutter button being pressed all the way down and the shutter release when shooting in hi-res shot mode. If the inteval is set, ▣ is displayed as an option for sequential shooting/self-timer mode. When not using high resolution shooting, set to [Off].
⚡Charge Time	Sets the charging time for flashes other than the dedicated flash.

- Electronic shutter is used for high resolution shooting.
- ☞ "Using a flash (Flash photography)" (P. 57)

Shooting with remote control wireless flash

Shooting with wireless flash is possible using the supplied flash unit with remote control wireless flash units. ☞ "Wireless remote control flash photography" (P. 153)

Using the Video Menu

Movie recording functions are set in the Video Menu.

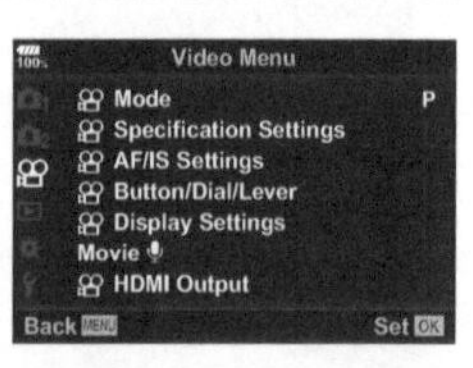

Option	Description	☞
Mode	Choose a movie record mode.	102
Specification Settings	Sets the image quality for movie recording. [ ]: Set a combination of image quality size and bit rate. [Noise Filter]: Select a noise reduction level for recording high-sensitivity movies. [Picture Mode]: Record in a picture mode suitable for editing when set to [On].	102
AF/IS Settings	[AF Mode]: Choose the AF mode for movie recording. [Image Stabilizer]: Sets image stabilization for movie recording.	43, 51, 53
Button/Dial/ Lever	Set the button, dial, and lever functions for movie mode. [Button Function]: Sets functions to buttons for movie mode. [Dial Function]: Sets the functions of the rear dial and front dial for movie mode. If [Exposure] is assigned to the front or rear dial, exposure compensation is possible up to ±3 EV (steps of 1/2 and 1 EV are also supported). [Fn Lever Function]: Sets functions to be switched by the **Fn** lever in movie mode. The **Fn** lever switches to the function set with [Dial Function] when set to [mode1]. If selecting [mode2] and pressing ▷, functions to be switched by the **Fn** lever can be selected from AF Mode, [/ / /] (AF target setting), and (AF area). This setting has no effect when [mode3] is selected for [Fn Lever Function] (P. 113) or when [Power 1] or [Power 2] is selected for [Fn Lever/Power Lever] (P. 113). [Shutter Function]: Sets the shutter button function for movie mode. When set to [ ], pressing the shutter button initiates autofocus. Still image photography is not available. When set to [◉ REC], movie recording can be started or stopped by pressing the shutter button all the way down. The ◉ button cannot be used for starting or stopping movie recording with this setting. [Elec. Zoom Speed]: Sets the zoom speed for the power zoom lens operation by the zoom ring.	—

Option	Description	☞
🎥 **Display Settings**	[🎥 Control Settings]: Sets whether to display Live Control (P. 126) and Live SCP (P. 50) in movie mode. For setting not to display, select the item and press the Ⓢ button to clear the check mark. [🎥 Info Settings]: Sets the information to be displayed on the movie recording screen. For setting not to display, select the item and press the Ⓢ button to clear the check mark. [Time Code Settings]: Set the time codes to record for movie mode. Set [Time Code Mode] to [Drop Frame] to record time codes corrected for errors with respect to recording time, and to [Non-DF] (no drop frame) to record uncorrected time codes. Set [Count Up] to [Rec Run] to run time codes during recording only, and to [Free Run] to run time codes even while recording is stopped, including when the camera is turned off. In [Starting Time], set a starting time for the time code. Set [Current Time] to set the time code for the current frame to 00. To set to 00:00:00:00, select [Reset]. You can also set time codes using [Manual Input]. Time codes cannot be recorded in Motion JPEG (HD). [🔋 Display Pattern]: Sets the battery level display ("%" or "minutes") in movie mode and during movie recording.	—
Movie 🎙	Audio will not be recorded in a movie when set to [Off].	103
🎥 **HDMI Output**	Sets the output setting for recording movies with this camera connected with an external device via HDMI. [Output Mode]: Sets the video output mode. When set to [Monitor Mode], image and camera information are output. The camera information is not displayed on the camera screen. When set to [Record Mode], only image is output. The camera information is displayed on the camera screen. [REC Bit]: If set to [On], the REC trigger is sent from the camera to the connected external device. [Time Code]: If set to [On], the time code is sent from the camera to the connected external device. • The recording by the external device that uses the time code as the REC trigger may be stopped in the following cases. - When recording an ART movie, etc., under heavy processing load - When switching the display between the monitor and viewfinder	—

Adding effects to a movie

You can create movies that take advantage of the effects available in still photography mode.

1 Select [Movie Mode] in Movie Video Menu and press the OK button.

2 Use △ ▽ to select an option and press the OK button.

P	Optimal aperture is set automatically according to the brightness of the subject. Use the front dial (front dial) or rear dial (rear dial) to adjust exposure compensation.
A	Depiction of background is changed by setting the aperture. Use the front dial (front dial) to adjust exposure compensation and rear dial (rear dial) to adjust aperture.
S	Shutter speed affects how the subject appears. Use the front dial (front dial) to adjust exposure compensation and rear dial (rear dial) to adjust shutter speed. Shutter speed can be set to values between 1/24 s and 1/8000 s.
M	Aperture and shutter speed can be manually set. Use the front dial (front dial) to select aperture value and the rear dial (rear dial) to select shutter speed. Shutter speed can be set to values between 1/3 s and 1/8000 s. ISO sensitivity can only be set manually to values between 200 and 6400.

- Blurred frames caused by such factors as the subject moving while the shutter is open can be reduced by selecting the fastest available shutter speed.
- The low end of the shutter speed changes according to the frame rate of the movie record mode.
- Excessive camera shake may not be compensated enough.
- When the inside of the camera becomes hot, shooting is automatically stopped to protect the camera.
- With some art filters, [C-AF] function is limited.

Setting the record mode (Movie record mode)

You can set combinations of movie image size and bit rate. The settings can be selected from options of movie record mode (P. 56).

1 Select [Movie Specification Settings] in Movie Video Menu and press the OK button.

2 Select [Movie record mode] and press ▷.

3 Use △ ▽ to select the item and press ▷.

- Use △ ▽ to select the setting and press the OK button.

Image size	Sets the image size to [C4K] (Custom only), [4K], [FHD] (Full HD), or [HD].
Bit rate	Sets the bit rate to [A-I] (All-Intra), [SF] (Super Fine), [F] (Fine), or [N] (Normal). • [A-I] (All-Intra) is not available for use with Clips. • When [4K] or [C4K] is set for the image size, you cannot select the bit rate.

Frame rate	Sets the frame rate to [60p], [50p], [30p], [25p], or [24p]. • [60p] and [50p] are not available in the following situations. - When [FHD] (Full HD) is set for the image size and [A-I] (All Intra) is set for the bit rate. - When [C4K] or [4K] is set for the image size. • If the image size is set to [C4K], the frame rate is locked to 24p.
Shooting time	Sets the shooting time to [8 sec], [4 sec], [2 sec], [1 sec], or [Off] (Custom only). The shooting time can only be set for Clips and Custom settings. • When [C4K] is set for the image size, the shooting time is locked to [Off].
Slow or Fast Motion	Sets slow or fast motion. Available settings differ depending on the set frame rate. • Slow and fast motions cannot be used in some image quality modes.

Setting the sound recording for movie (Movie 🎙)

Configure the sound recording settings for movie recording.

1 Select [Movie 🎙] in 🎥 Video Menu and press the Ⓞ button.

2 Select [On] and press ▷.

3 Use △▽ to select the item and press ▷.

- Use △▽ to select the setting and press the Ⓞ button.

Recording Volume	Adjust microphone sensitivity for the built-in microphone and for optional external microphones. Adjust the sensitivity using △▽ while checking the peak sound level picked up by the microphone over the previous few seconds.
🎙 Volume Limiter	If set to [On], the volume is automatically regulated when the volume picked up by the microphone is louder than normal.
Wind Noise Reduction	Reduce wind noise during recording.
🎙 Plug-in Power	Set to [On] for distributing power to the microphone, and [Off] for using a professional microphone, etc, which does not require power distribution from the camera.
PCM Recorder 🎙 Link	Set to [On] for using an IC recorder connected to the microphone jack as a microphone. ☞ "Recording movie audio with an IC recorder" (P. 104)
Headphone Volume	Set the volume for the connected headphone.

- Operation sounds of the lens and camera may be recorded in a movie. To prevent it from recording, reduce the operation sounds by setting [AF Mode] to [S-AF] or [MF], or by minimizing the button operations of the camera.
- Sound cannot be recorded in ART 7 (Diorama) mode.
- When [Movie 🎙] is set to [Off], 🎙OFF is displayed.

4

Menu functions (Video menu)

Recording movie audio with an IC recorder

You can use an IC recorder to record audio in a movie.
Connect an IC recorder to the microphone jack of the camera for recording sound. Use a non-resistant cable for the connection.

1. Select [Movie 🎙] in 🎥 Video Menu and press the ⓞⓚ button.
2. Select [On] and press ▷.
3. Use △▽ to select [PCM Recorder 🎙 Link] and press ▷.
4. Use △▽ to select the item and press ▷.
   - Use △▽ to select the setting and press the ⓞⓚ button.

Camera Rec. Volume	If set to [Inoperative], the sound recording settings in the camera are disabled and the settings in the IC recorder are applied.
Slate Tone	If set to [On], the slate tone is played.
Synchronized ◉ Rec.	If set to [On], the IC recorder automatically starts/ends sound recording at the same time when the movie recording starts/ends in the camera.

Recording movie audio with the Olympus LS-100 IC recorder

When using the Olympus LS-100 IC recorder to record sound in a movie, you can add a slate tone, start and stop recording with camera controls.
In [Movie 🎙] > [PCM Recorder 🎙 Link], set [Slate Tone] and [Synchronized ◉ Rec.] to [On].
Make sure that the LS-100 firmware is the latest version before recording.

1. Connect the LS-100 to the USB connector and microphone.
   - When the LS-100 is connected to the USB connector, a message will be displayed prompting you to choose a connection type. Select [PCM Recorder]. If the message does not appear, select [Auto] for [USB Mode] (P. 117) in the custom menus.
2. Start recording a movie.
   - Audio recording on the LS-100 starts simultaneously.
   - If you press and hold the ⓞⓚ button, you can record a slate tone.
3. End movie recording.
   - Audio recording on the LS-100 ends simultaneously.

- Refer to the documentation provided with the LS-100 as well.

Using the Playback Menu

Playback Menu

(P. 80)
(P. 105)
Edit (P. 105)
Print Order (P. 144)
Reset Protect (P. 108)
Copy All (P. 108)
Connection to Smartphone (P. 135)

Displaying images rotated ()

If set to [On], images in portrait orientation are automatically rotated to be displayed in the correct orientation on the playback display.

Editing still images

Recorded images can be edited and saved as separate images.

- The images to be edited can be chosen from those on the card currently selected for playback. ☞ "Setting the card to record to" (P. 132)

1 Select [Edit] in the ▶ Playback Menu and press the OK button.

2 Use △▽ to select [Sel. Image] and press the OK button.

3 Use ◁▷ to select the image to be edited and press the OK button.

- [RAW Data Edit] is displayed if a RAW image is selected, and [JPEG Edit] if a JPEG image is selected. For images recorded in RAW+JPEG format, both [RAW Data Edit] and [JPEG Edit] are displayed. Select the desired option from them.

4 Select [RAW Data Edit] or [JPEG Edit] and press the OK button.

RAW Data Edit	Creates a JPEG copy of a RAW image according to the selected settings.	
	Current	The JPEG copy is processed using the current camera settings. Adjust the camera settings before choosing this option. Some settings such as exposure compensation are not applied.
	Custom1	Edit by changing the settings on the screen. The used settings can be saved.
	Custom2	
	ART BKT	The image is edited using settings for the selected art filter.

JPEG Edit	Choose from the following options: [Shadow Adj]: Brightens a dark backlit subject. [Redeye Fix]: Reduces the red-eye phenomenon due to flash shooting. [⌗]: Trims an image. Use the front dial (⊙) or rear dial (⊙) to choose the size of the trimming and △▽◁▷ to specify the trimming position. 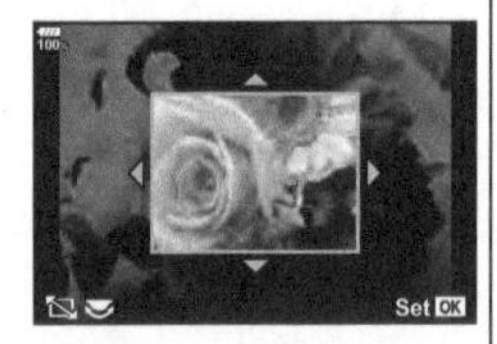 [Aspect]: Changes the aspect ratio of images from 4:3 (standard) to [3:2], [16:9], [1:1], or [3:4]. After changing the aspect ratio, use △▽◁▷ to specify the trimming position. [Black & White]: Creates black and white images. [Sepia]: Creates sepia-toned images. [Saturation]: Increases the vividness of images. Adjust the color saturation checking the image on the screen. [⊡]: Converts the image size to 1280 × 960, 640 × 480, or 320 × 240. Images with an aspect ratio other than 4:3 (standard) are converted to the closest image size. [e-Portrait]: Compensates the skin look for smoothness. Compensation cannot be applied in such a case that a face cannot be detected.

5 When the settings are complete, press the Ⓞ button.

- The settings are applied to the image.

6 Select [Yes] and press the Ⓞ button.

- The edited image is stored in the card.

- Red-eye correction may not work depending on the image.
- Editing of a JPEG image is not possible in the following cases:
  When an image is processed on a PC, when there is not enough space in the card memory, or when an image is recorded on another camera.
- The image cannot be resized (⊡) to the larger size than the original size.
- [⌗] (trimming) and [Aspect] can only be used to edit images with an aspect ratio of 4:3 (standard).
- When [ART] is selected for picture mode, [Color Space] (P. 65) will be locked at [sRGB].

Image overlay

Up to 3 frames of RAW images taken with the camera can be overlaid and saved as a separate image.
The image is saved with the record mode set at the time the image is saved. (If [RAW] is selected, the copy will be saved in [LN+RAW] format.)

1 Select [Edit] in the ▶ Playback Menu and press the (OK) button.

2 Use △▽ to select [Image Overlay] and press the (OK) button.

3 Select the number of images to be overlaid and press the (OK) button.

4 Use △▽◁▷ to select the RAW images to be overlaid.
- The overlaid image will be displayed if images of the number specified in step 3 are selected.

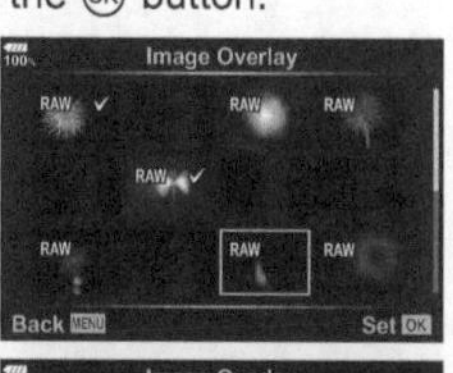

5 Adjust the gain for each image to be overlaid.
- Use ◁▷ to select an image and △▽ to adjust gain.
- Gain can be adjusted in the range 0.1 – 2.0. Check the results in the monitor.

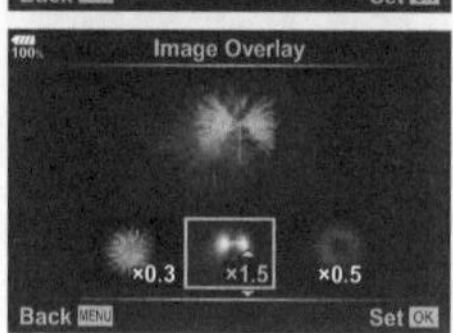

6 Press the (OK) button to display the confirmation dialog. Select [Yes] and press the (OK) button.

Tips
- To overlay 4 or more frames, save the overlay image as a RAW file and use [Image Overlay] repeatedly.

Audio recording

Audio can be added to still images (up to 30 sec. long).
This is the same function as [🎙] during playback (P. 83).

Saving a still image from a movie (In-Movie Image Capture)

You can select a frame from a movie to save as a still image.

1 Select [Edit] in the ▶ Playback Menu and press the (OK) button.

2 Use △▽ to select [Sel. Image] and press the (OK) button.

3 Use ◁▷ to select a movie and press the (OK) button.

4 Select [Movie Edit] and press the (OK) button.

5 Use △▽ to select [In-Movie Image Capture] and press the (OK) button.

6 Use ◁▷ to select a frame to be saved as a still image and press the (OK) button.

- Movies recorded by this camera with the aspect ratio set to [16:9] and image quality mode set to 4K size in MOV can be edited.

Trimming movies (Movie Trimming)

1 Select [Edit] in the ▶ Playback Menu and press the ⊛ button.
2 Use △▽ to select [Sel. Image] and press the ⊛ button.
3 Use ◁▷ to select a movie and press the ⊛ button.
4 Select [Movie Edit] and press the ⊛ button.
5 Use △▽ to select [Movie Trimming] and press the ⊛ button.
6 Select [Overwrite] or [New File] and press the ⊛ button.
   - If the image is protected, you cannot select [Overwrite].
7 Specify an area to trim.
   - The range between the first or last frame and the selected frame are deleted.
8 Select [Yes] and press the ⊛ button.

- Editing is available with movies recorded using this camera.

Canceling all protections

Protections of multiple images can be canceled at a time.

1 Select [Reset Protect] in the ▶ Playback Menu and press the ⊛ button.
2 Select [Yes] and press the ⊛ button.
   - All the protections of images saved in the card being played back will be canceled.

Copy All

All images can be copied between the cards inserted in the camera (card slot 1 and 2).

1 Select [Copy All] in the ▶ Playback Menu and press the ⊛ button.
2 Select an option and press the ⊛ button.

1⇨2	All images are copied from the card in card slot 1 to the card in card slot 2.
2⇨1	All images are copied from the card in card slot 2 to the card in card slot 1.

3 Select [Yes] and press the ⊛ button.

- Copying ends when the destination card is full.

Using the setup menu

Use the Setup Menu to set the basic camera functions.

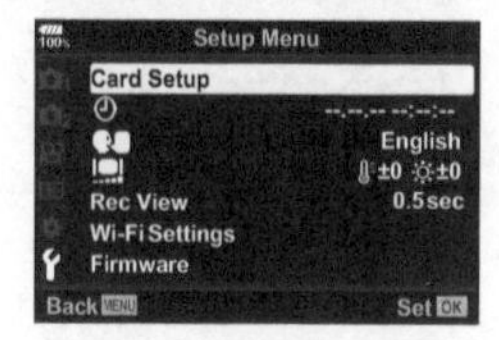

Option	Description	☞
Card Setup	Format the card and delete all images.	110
(Date/time setting)	Set the camera clock.	19
(Changing the display language)	You can change the language used for the on-screen display and error messages.	—
(Monitor brightness adjustment)	You can adjust the brightness and color temperature of the monitor. Color temperature adjustment is only applied to the monitor display during playback. Use ◁▷ to highlight (color temperature) or (brightness) and △▽ to adjust the value. Press the **INFO** button to switch the saturation of the monitor between [Natural] and [Vivid] settings.	—
Rec View	Sets whether to display the captured image on the monitor after shooting, and the length of time for the display. This is useful for a brief check of the picture you have taken. You can shoot the next shot by pressing the shutter button halfway even while the captured image is displayed on the monitor. [0.3sec]–[20sec]: Sets the length of time (seconds) to display the captured image on the monitor. [Off]: The captured image is not displayed on the monitor. [AUTO ▶]: Displays the captured image, and then switches to playback mode. This is useful for erasing a picture after checking it.	—
Wi-Fi Settings	Sets the wireless connection method for the camera to connect with smartphones that support wireless LAN connections.	137
Firmware	Displays the firmware versions of the camera and connected accessories. Check the versions when you inquire about the camera or accessories or when you download the software.	—

Formatting the card (Card Setup)

Cards must be formatted with this camera before first use or after being used with other cameras or computers.
All data stored on the card, including protected images, is erased when the card is formatted.
When formatting a used card, confirm there are no images that you still want to keep on the card. ☞ "Usable cards" (P. 147)

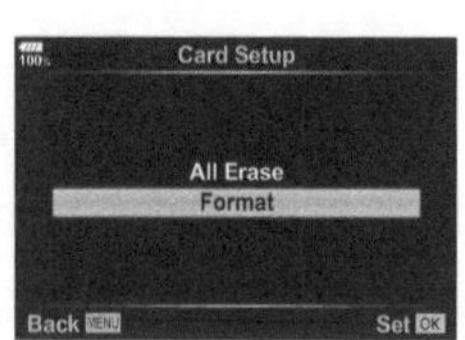

1 Select [Card Setup] in the Setup Menu and press the OK button.
- When there are cards in both slots 1 and 2, card slot selection appears. Select a card slot and press the OK button.
- If there is data on the card, menu items appear. Select [Format] and press the OK button.

2 Select [Yes] and press the OK button.
- Formatting is performed.

Deleting all images (Card Setup)

All images on a card can be deleted at a time. Protected images are not deleted.

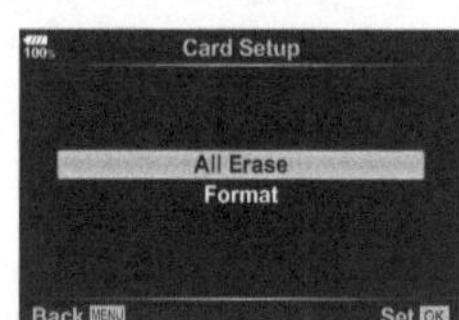

1 Select [Card Setup] in the Setup Menu and press the OK button.
- When there are cards in both slots 1 and 2, card slot selection appears. Select a card slot and press the OK button.

2 Select [All Erase] and press the OK button.

3 Select [Yes] and press the OK button.
- All images are deleted.

Using the custom menus

Camera settings can be customized using the ✲ Custom Menu.

Custom Menu

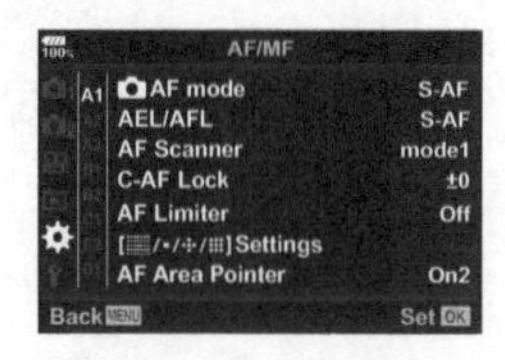

A1 AF/MF

MENU ➔ ✲ ➔ A1

Option	Description	☞
◘ AF Mode	Choose the AF mode for still image shooting.	43, 51
AEL/AFL	Customize AF and AE lock.	123
AF Scanner	Sets the AF scan function* for when the camera cannot focus on the subject or the contrast is not clear. * Scans whole range from minimum range to infinity for focus point when the camera cannot focus on the subject or the contrast is not clear. [mode1]: AF Scanner is not activated. [mode2]: AF Scanner is activated once only. [mode3]: AF Scanner is activated.	—
C-AF Lock	Sets the tracking sensitivity for C-AF.	—
AF Limiter	Limits the AF area when turned [On]. [Distance settings]: You can register the AF Limiter area. You can register up to 3 distance areas. Numerical value and unit (m, ft) can be set. The distance is approximate and not exact. [Pri. Rls]: If [On] is selected, the shutter can be released while AF Limiter is operating even when the camera is not in focus. • AF Limiter is not available in the following cases. - When the focus limiter is enabled on the lens. - When using focus bracketing - While in the movie mode or recording a movie	—
[▦/•/✣/▦] Settings	Sets functions that appear in AF Target settings. To hide an item, select the item and press the Ⓞ button to clear the check mark.	39
AF Area Pointer	[On1]: Displays the AF target frame in green. [On2]: Displays the AF target frame in green while the shutter button is pressed halfway. If you select [Off], the AF target frame will not be displayed during confirmation.	—

A2 AF/MF

MENU ➜ ✲ ➜ A2

Option	Description	☞
AF Targeting Pad	If [On] is selected, the AF target can be positioned by tapping the monitor during viewfinder photography. Tap the monitor and slide your finger to position the AF target. • When set to [On], drag operation can be disabled or enabled by double-tapping the monitor. • [AF Targeting Pad] can also be used with zoom frame AF (P. 41).	—
[·:·] Set Home	Set the AF target mode, AF target position, and AF mode that are used as the home position. Press Ⓞ button to select the desired options in the [[·:·] Set Home] display. HP appears in the AF target selection display while you choose a home position.	—
[·:·] Custom Settings	You can change the dial and △▽◁▷ button functions for the AF Area screen. • To use the settings stored in [Set 2], highlight [Set 2] in the [[·:·] Custom Settings] menu and press Ⓞ button. • You can switch to [Set 2] by pressing the **INFO** button in the AF-target selection display.	—
AF Illuminator	Select [Off] to disable the AF illuminator.	—
☺ Face Priority	You can select the face priority or eye priority AF mode.	40
AF Focus Adj.	Focal location adjustment for phase-difference AF can be fine tuned within a range of ±20 steps.	—

A3 AF/MF

MENU ➜ ✲ ➜ A3

Option	Description	☞
Preset MF distance	Sets the Preset MF focus position. Numerical value and unit (m, ft) can be set. The distance is approximate and not exact.	—
MF Assist	When set to [On], you can automatically switch to zoom or peaking in manual focus mode by rotating the focus ring.	124
MF Clutch	Selecting [Inoperative] prevents the lens MF clutch and snapshot focus being used for manual focus. To focus manually, slide the focus ring forward.	—
Focus Ring	You can customize how the lens adjusts to the focal point by selecting the rotational direction of the focus ring.	—
Bulb/Time Focusing	You can change the focus position during exposure by using manual focus (MF). When set to [Off], rotation of the focus ring is disabled.	—
Reset Lens	When set to [Off], the focus position of the lens is not reset even when the power is turned off. When set to [On], the focus of power zoom lenses is also reset.	—

B Button/Dial/Lever

MENU → ✲ → B

Option	Description	☞
Button Function	Choose the function assigned to the selected button.	66
PBH Lock	If [On] is selected, △▽◁▷ and ⊛ button operations for PBH (Power Battery Holder) are disabled.	—
Dial Function	You can change the function of the front dial and rear dial.	—
Dial Direction	Choose the direction in which the dial is rotated to adjust shutter speed or aperture. Change the program shift direction in which the dial is rotated.	—
Fn Lever Settings	[Fn Lever Function]: You can switch the dial and button function depending on the position of the **Fn** lever. [Switch ● Function]: When turned [On], you can switch the ● button function depending on the position of the **Fn** lever. When [On] is selected and the **Fn** lever is set to position 2, the **AF** button activates the flash and the **HDR** button switches the setting to bracketing setting.	124
Fn Lever/Power Lever	[Fn]: Follows settings for the **Fn** lever function. [Power 1]: The power turns on when the **Fn** lever is in position 1, and off when in position 2. [Power 2]: The power turns on when the **Fn** lever is in position 2, and off when in position 1. When [Power 1] or [Power 2] is set, the **ON/OFF** lever (power lever), [Fn Lever Settings] and [Fn Lever Function] are disabled.	—
Elec. Zoom Speed	You can change the zoom speed used when operating the power zoom lens with the zoom ring during still image shooting.	—

C1 Release/▣/Image Stabilizer

MENU ➜ ✲ ➜ C1

Option	Description	☞
Rls Priority S Rls Priority C	If [On] is selected, the shutter can be released even when the camera is not in focus. This option can be set separately for S-AF and C-AF modes (P. 43, 51).	—
▣L Settings ▣H Settings	You can select from [▣], [♦▣], [♥▣], and [Pro Cap] for the sequential shooting speed and shot limit. You can set the pre-shutter frames for [Pro Cap]. Figures for sequential shooting speed are the approximate maximums.	48

C2 Release/▣/Image Stabilizer

MENU ➜ ✲ ➜ C2

Option	Description	☞
📷 Image Stabilizer	Sets image stabilization for still image shooting.	53
▣ Image Stabilization	Sets the priority function during sequential shooting. [Fps Priority]: Shooting speed gets priority over image stabilization. The sensor will not be reset to the center during sequential shooting. [IS Priority]: Image stabilization gets priority over shooting speed. The sensor will be reset to the center per frame of sequential shooting. The shooting speed will drop slightly.	—
Half Way Rls With IS	When set to [Off], the IS (Image Stabilization) function while the shutter button is pressed halfway will not be activated.	—
Lens I.S. Priority	If [On] is selected, priority is given to the lens function operation when using a lens with an image stabilization function. • This option has no effect on lenses that are equipped with an image stabilization switch.	—

D1 Disp/■)))/PC

MENU ➜ ✲ ➜ D1

Option	Description	☞
Control Settings	Choose the controls displayed in each shooting mode. <table><tr><th rowspan="2">Controls</th><th colspan="3">Shooting mode</th></tr><tr><th>iAUTO</th><th>P/A/S/M</th><th>ART</th></tr><tr><td>Live Control (P. 126)</td><td>✓</td><td>✓</td><td>✓</td></tr><tr><td>Live SCP (P. 50)</td><td>✓</td><td>✓</td><td>✓</td></tr><tr><td>Live Guide (P. 31)</td><td>✓</td><td>–</td><td>–</td></tr><tr><td>Art Menu (P. 33)</td><td>–</td><td>–</td><td>✓</td></tr></table>Press the **INFO** button to switch the screen contents.	125
/Info Settings	Choose the information displayed when the **INFO** button is pressed. [▶ Info]: Choose the information displayed in full frame playback. [▶Q Info]: Choose the information displayed in magnified playback. [LV-Info]: Choose the information displayed when the camera is in shooting mode. [Settings]: Choose the information displayed in index, "My Clips", and calendar playback.	127, 128
Picture Mode Settings	Select a function to display in the picture mode type selection screen (P. 61). To hide an item, select the item and press the Ⓚ button to clear the check mark.	—
/ Settings	Select a function to display in the sequential shooting/self-timer function selection screen (P. 46). To hide an item, select the item and press the Ⓚ button to clear the check mark.	—
Multi Function Settings	Select a multi function (P. 70) option. When not using this option, select the item and press the Ⓚ button to clear the check mark.	—

D2 Disp/■)))/PC

MENU ➜ ✲ ➜ D2

Option	Description	☞
Live View Boost	Shoot while checking the subject even under low-light conditions. In **M** mode, you can use this setting when shooting with BULB/TIME shooting and live composite. [On1]: Prioritize smoothness of display. [On2]: Prioritize image visibility in dark conditions.	—
Art LV Mode	[mode1]: The filter effect is always displayed. [mode2]: Priority is given to smooth display while the shutter button is pressed halfway. The quality of art filter effect previews may be affected.	—
Frame Rate	If set to [High], a moving subject can be tracked more smoothly. However, the number of frames to be shot will decrease slightly. This setting is automatically set to [Standard] if the camera becomes hot.	—

D2 Disp/■)))/PC

MENU ➔ ✲ ➔ D2

Option	Description	☞
LV Close Up Settings	[LV Close Up Mode]: When set to [mode1], pressing the button halfway in the magnified live view returns to the magnified frame display. When set to [mode2], pressing the button halfway in the magnified live view switches to the Zoom AF display. [Live View Boost]: When set to [On], the magnified area is adjusted for proper exposure. This is useful to check the focus when shooting in dark locations. When set to [Off], the magnified area is displayed with the brightness of live view before magnification. This is useful to check the focus when shooting in backlit locations.	—
⊙ Settings	[⊙ Lock]: Select [On] to maintain the aperture at the selected value even when releasing the button. [Live View Boost]: Select [On] to shoot while checking the subject even under low-light conditions.	—
Flicker reduction	Reduce the effects of flicker under some kinds of lighting, including fluorescent lamps. When flicker is not reduced by the [Auto] setting, set to [50Hz] or [60Hz] in accordance with the commercial power frequency of the region where the camera is used.	—

D3 Disp/■)))/PC

MENU ➔ ✲ ➔ D3

Option	Description	☞
Grid Settings	Sets the guide line display that appears when shooting. [Display Color]: Sets the guide line color and opacity. Can be set to [Preset 1] and [Preset 2]. [Displayed Grid]: Select [▦], [⊞], [⊞], [⊠], [▭] or [⊞] to display a grid on the monitor. [Apply Settings to EVF]: If [On] is selected, the guides shown in the monitor will also be displayed in the viewfinder when [Style 1] or [Style 2] is selected for [EVF Style]. The setting selected in [EVF Grid Settings] will be invalid.	—
Peaking Settings	You can change the edge enhancement color and intensity. The edge enhancement color (red, yellow, white, black) and intensity (Standard, Low, High), and the brightness of the peaking background (On, Off) can be set. • If [Image Brightness Adj.] is set to [On], the brightness of live view is adjusted to enhance the enhancement color.	124
Histogram Settings	[Highlight]: Choose the lower bound for the highlight display. [Shadow]: Choose the upper bound for the shadow display.	127
Mode Guide	Select [On] to display a help for the selected mode when the mode dial is rotated to a new setting.	24
Selfie Assist	Selecting [On] optimizes the display for self-portraits when the monitor is in the self-portrait position.	129

D4 Disp/■)))/PC

MENU ➔ ✲ ➔ D4

Option	Description	☞
■))) (Beep sound)	When set to [Off], you can turn off the beep sound that is emitted when the focus locks by pressing the shutter button.	—
HDMI	[Output Size]: Selecting the digital video signal format for connecting to a TV via an HDMI cable. [HDMI Control]: Select [On] to allow the camera to be operated using remotes for TVs that support HDMI control. This option takes effect when pictures are displayed on a TV. [Output Frame Rate]: Select the output frame rate from [50p Priority] or [60p Priority] for using the camera connected to a TV with an HDMI cable.	130
USB Mode	Choose a mode for connecting the camera to a computer or printer. Choose [Auto] to display USB mode options every time the camera is connected. When you select [🖳/📷], you can use dedicated software to control the camera from a computer and transfer images from the camera to a computer. Access the following URL to download and install the dedicated software. To use [🖳/📷], you must first set the mode dial to **P**, **A**, **S**, or **M** mode. http://support.olympus-imaging.com/oc1download/index/	—

E1 Exp/ISO/BULB/▣

MENU ➔ ✲ ➔ E1

Option	Description	☞
Exposure Shift	Adjust correct exposure separately for each metering mode. • This reduces the number of exposure compensation options available in the selected direction. • The effects are not visible in the monitor. To make normal adjustments to the exposure, perform exposure compensation (P. 39).	—
EV Step	Choose the size of the increments used when selecting shutter speed, aperture, exposure compensation, and other exposure parameters.	—
ISO Step	Select the increments available for choosing ISO sensitivity.	—
ISO-Auto Set	[Upper Limit / Default]: Choose the upper limit and default value used for ISO sensitivity when [Auto] is selected for ISO. [High Limit]: Choose the upper limit for auto ISO sensitivity selection. [Default]: Choose the default value for auto ISO sensitivity selection. The maximum is 6400. [Lowest S/S Setting]: Automatically sets the shutter speed when the ISO sensitivity is raised in **P** and **A** modes. If set to [Auto], the camera automatically sets the shutter speed.	—
ISO-Auto	Choose the shooting modes in which [Auto] ISO sensitivity is available. [**P**/**A**/**S**]: Auto ISO sensitivity selection is available in all modes except **M**. [All]: Auto ISO sensitivity selection is available in all modes.	—

E1 Exp/ISO/BULB/ MENU ➜ ✲ ➜ E1

Option	Description	☞
Noise Filter	Choose the amount of noise reduction performed at high ISO sensitivities.	—
Noise Reduct.	This function reduces the noise that is generated during long exposures. [Auto]: Noise reduction is performed at slow shutter speeds, or when the internal temperature of the camera has risen. [On]: Noise reduction is performed with every shot. [Off]: Noise reduction off. • The time required for noise reduction is shown in the display. • [Off] is selected automatically during sequential shooting. • This function may not work effectively with some shooting conditions or subjects.	29

E2 Exp/ISO/BULB/ MENU ➜ ✲ ➜ E2

Option	Description	☞
Bulb/Time Timer	Choose the maximum exposure for bulb and time photography.	—
Bulb/Time Monitor	Set the monitor brightness when [BULB], [TIME], or [Live Composite] is used.	—
Live Bulb	Choose the display interval during shooting. The number of update times is limited. The frequency drops at high ISO sensitivities. Choose [Off] to disable the display. Tap the monitor or press the shutter button halfway to refresh the display.	—
Live Time		—
Composite Settings	Set an exposure time to be the reference in composite photography .	30

E3 Exp/ISO/BULB/ MENU ➜ ✲ ➜ E3

Option	Description	☞
Metering	Choose a metering mode according to the scene.	45, 51
AEL Metering	Choose the metering method used for AE lock (P. 45). [Auto]: Use the currently selected metering method.	—
[·:·] Spot Metering	Choose whether the [Spot], [Spot Hilight], and [Spot Shadow] spot metering options meter the selected AF target.	—

F ϟ Custom MENU ➜ ✲ ➜ F

Option	Description	☞
ϟ X-Sync.	Choose the shutter speed used when the flash fires.	131
ϟ Slow Limit	Choose the slowest shutter speed available when a flash is used.	131

F ϟ Custom

MENU ➔ ✲ ➔ F

Option	Description	☞
ϟ±+±	When set to [On], the exposure compensation value is added to the flash compensation value.	39, 60
ϟ+WB	Adjust white balance for use with a flash.	—

G ◀:-/WB/Color

MENU ➔ ✲ ➔ G

Option	Description	☞
◀: Set	You can select the JPEG image quality mode from combinations of three image sizes and four compression rates. 1) Use ◁▷ to select a combination ([◀:-1] – [◀:-4]) and use △▽ to change. 2) Press the Ⓞ button. Image size / Compression rate	55, 88, 131
Pixel Count	Choose the pixel count for [M]- and [S]-size images. 1) Select [Middle] or [Small] and press ▷. 2) Choose a pixel count and press the Ⓞ button.	55, 88, 131
Shading Comp.	Choose [On] to correct peripheral illumination according to the type of lens. • Compensation is not available for teleconverters or extension tubes. • Noise may be visible at the edges of photographs taken at high ISO sensitivities.	—
WB	Set the white balance. You can also fine-tune the white balance for each mode.	42, 52
All WB±	[All Set]: Use the same white balance compensation in all modes except [CWB]. [All Reset]: Set white balance compensation for all modes except [CWB] to 0.	—
WB AUTO Keep Warm Color	Select [On] to preserve “warm” colors in pictures taken under incandescent lighting.	—
Color Space	You can select a format to ensure that colors are correctly reproduced when shot images are regenerated on a monitor or using a printer.	65

H1 Record/Erase

MENU ➜ ✲ ➜ H1

Option	Description	☞
Card Slot Settings	Sets the card for recording still images or movies.	132
File Name	[Auto]: Even when a new card is inserted, the file numbers are retained from the previous card. File numbering continues from the last number used or from the highest number available on the card. [Reset]: When you insert a new card, the folder numbers starts at 100 and the file name starts at 0001. If a card containing images is inserted, the file numbers start at the number following the highest file number on the card. When recording data to two cards at the same time, files are numbered under the same rule as for the single card according to the file number and folder number of both cards.	—
Edit Filename	Choose how image files are named by editing the portion of the filename highlighted below in gray. sRGB: Pmdd0000.jpg —— Pmdd Adobe RGB: _mdd0000.jpg —— mdd	—
dpi Settings	Choose the print resolution.	—
Copyright Settings*	Add the names of the photographer and copyright holder to new photographs. Names can be up to 63 characters long. [Copyright Info.]: Select [On] to include the names of the photographer and copyright holder in the Exif data for new photographs. [Artist Name]: Enter the name of the photographer. [Copyright Name]: Enter the name of the copyright holder. **1)** Select characters from ① and press the Ⓞ button. The selected characters appear in ②. **2)** Repeat Step 1 to complete the name, then highlight [END] and press the Ⓞ button. • To delete a character, press the **INFO** button to place the cursor in the name area ②, highlight the character, and press 🗑. ② Copyright Name 05/63 ABCD ① Cancel Delete Set * OLYMPUS does not accept liability for damages arising from disputes involving the use of [Copyright Settings]. Use at your own risk.	—
Lens Info Settings	Save lens info for up to 10 lenses that do not automatically supply info to the camera.	132

H2 Record/Erase

MENU ➔ ✲ ➔ H2

Option	Description	☞
Quick Erase	If [On] is selected, pressing the 🗑 button in the playback display will immediately delete the current image.	—
RAW+JPEG Erase	Choose the action performed when a photograph recorded at a setting of RAW+JPEG is erased in single-frame playback . [JPEG]: Only the JPEG copy is erased. [RAW]: Only the RAW copy is erased. [RAW+JPEG]: Both copies are erased. • Both the RAW and JPEG copies are deleted when selected images are deleted or when [All Erase] (P. 110) is selected.	55, 82, 88
Priority Set	Choose the default selection ([Yes] or [No]) for confirmation dialogs.	—

I EVF

MENU ➔ ✲ ➔ I

Option	Description	☞
EVF Auto Switch	If [Off] is selected, the viewfinder will not turn on when you put your eye to the viewfinder. Use the [◯] button to choose the display.	—
EVF Adjust	Adjust viewfinder brightness and hue. Brightness is automatically adjusted when [EVF Auto Luminance] is set to [On]. The contrast of the information display is also adjusted automatically.	—
EVF Style	Choose the viewfinder display style.	133
▭ Info Settings	Like the monitor, the viewfinder can be used to display histograms, highlights and shadows, and level gauge. The level gauge is available when [Style 1] or [Style 2] is set in [EVF Style].	—
EVF Grid Settings	Choose the type and color of framing grid displayed in the viewfinder when [Off] is selected for [Apply Settings to EVF] and [Style 1] or [Style 2] is selected for [EVF Style]. Choose the framing grid from [▦], [⊞], [⊞], [⊠], [▤] or [⊞].	—
▭ Half Way Level	If set to [Off], the level gauge will not be displayed when the shutter button is pressed halfway down. The level gauge is available when [Style 1] or [Style 2] is set in [EVF Style].	—
S-OVF	Select [On] for a viewfinder display similar to an optical viewfinder. Selecting [S-OVF] makes the details in shadows easier to see. • S-OVF is displayed in the viewfinder when [S-OVF] starts. • The display is not adjusted for settings such as white balance, exposure compensation, and picture mode.	—

J1 Utility

Option	Description	☞
Pixel Mapping	The pixel mapping feature allows the camera to check and adjust the image pickup device and image processing functions.	159
Press-and-hold Time	Set the press-and-hold time until the function assigned to the button operates, from [0.5 sec] to [3.0 sec].	—
Level Adjust	You can calibrate the angle of the level gauge. [Reset]: Resets adjusted values to the default settings. [Adjust]: Sets the current camera orientation as the 0 position.	—
Touchscreen Settings	Activate the touch screen. Choose [Off] to disable the touch screen.	—
Menu Recall	Set [Recall] to display the cursor at the last position of operation when you display a menu. The cursor position will be retained even when you turn off the camera.	—

J2 Utility

MENU ➜ ✲ ➜ J2

Option	Description	☞
Battery Settings	[Battery Priority]: Sets the preferred battery to use. When [Body Battery] is selected, the battery in the body takes priority. When [PBH Battery] is selected, the battery in the PBH (Power Battery Holder) takes priority. [Battery Status]: Displays the status of the equipped battery.	—
Backlit LCD	If no operations are performed for the selected period, the backlight will dim to save battery power. The backlight will not dim if [Hold] is selected.	—
Sleep	The camera will enter sleep (energy saving) mode if no operations are performed for the selected period. The camera can be reactivated by pressing the shutter button halfway.	—
Auto Power Off	When in sleep mode, the camera will automatically turn off after the set time has elapsed.	—
Quick Sleep Mode	When set to [On], the camera will go into energy-saving mode during shooting without using live view, allowing the camera to take pictures while using less power. The backlight time and sleep time can be set. Press the shutter button to return from energy-saving mode. The camera will not go into energy-saving mode while the live view is displayed or the viewfinder is in use. The ECO icon is displayed on the super control panel when set to [On].	—
Eye-Fi*	Enable or disable upload when using an Eye-Fi card. The setting can be changed when an Eye-Fi card is inserted.	—
Certification	Display certification icons.	—

* Use the Eye-Fi card in compliance with the laws and regulations of the country where the camera is used. Onboard airplanes and in other locations in which the use of wireless devices is prohibited, remove the Eye-Fi card from the camera, or select [Off] for [Eye-Fi]. The camera does not support the "endless" Eye-Fi mode.

AEL/AFL

MENU ➜ ✲ ➜ A1 ➜ [AEL/AFL]

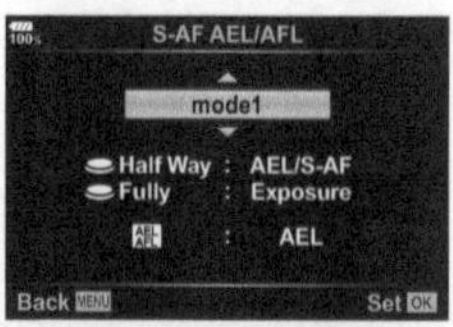

Autofocus and metering can be performed by pressing the button to which AEL/AFL has been assigned. Choose a mode for each focus mode.

Assignments of AEL/AFL function

Mode		Shutter button function				AEL/AFL Button function	
		Half-press		Full press		When holding down AEL/AFL	
		Focus	Exposure	Focus	Exposure	Focus	Exposure
S-AF	mode1	S-AF	Locked	–	–	–	Locked
	mode2	S-AF	–	–	Locked	–	Locked
	mode3	–	Locked	–	–	S-AF	–
C-AF	mode1	C-AF start	Locked	Locked	–	–	Locked
	mode2	C-AF start	–	Locked	Locked	–	Locked
	mode3	–	Locked	Locked	–	C-AF start	–
	mode4	–	–	Locked	Locked	C-AF start	–
MF	mode1	–	Locked	–	–	–	Locked
	mode2	–	–	–	Locked	–	Locked
	mode3	–	Locked	–	–	S-AF	–

MF Assist

MENU ➔ ✲ ➔ A3 ➔ [MF Assist]

This is a focus assist function for MF. When the focus ring is rotated, the edge of the subject is enhanced or a portion of the screen display is magnified. When you stop operating the focus ring, the screen returns to the original display.

Magnify	Magnifies a portion of the screen. The portion to be magnified can be set in advance using AF target. ☞ "Setting the AF target" (P. 40)
Peaking	Displays clearly defined outlines with edge enhancement. You can select the enhancement color and intensity. ☞ [Peaking Settings] (P. 116)

- [Peaking] can be displayed using button operations. The display is switched every time the button is pressed. Assign the switching function to one of the buttons in advance using Button Function (P. 66).
- Press the **INFO** button to change the color and intensity when Peaking is displayed.
- When Peaking is in use, the edges of small subjects tend to be enhanced more strongly. This is no guarantee of accurate focusing.

Fn Lever Function

MENU ➔ ✲ ➔ B ➔ [Fn Lever Settings] ➔ [Fn Lever Function]

You can switch the dial and button function depending on the position of the **Fn** lever.

Fn lever position and dial/button function list

Mode	**Fn lever position 1**	**Fn lever position 2**
Off	The **Fn** lever function is turned off.	
mode1	The dial function operates according to settings in [Dial Function] (P. 113).	
mode2	Sets functions to be switched by the **Fn** lever from AF Mode, [▦/▪/⁘/▦] (AF target setting), and ⊕ (AF area).	
mode3	Sets the shooting mode selected by the mode dial.	Switches to the movie mode.

Choosing the control panel displays ([camera icon]Control Settings)

MENU ➜ ✲ ➜ D1 ➜ [[camera icon]Control Settings]

Sets whether or not to display control panels for option selection in each shooting mode.

In each shooting mode, press the Ⓞ button to insert a check in the control panel you want to display.

How to display control panels

- Press the Ⓞ button while the control panel is displayed, then press the **INFO** button to switch the display.
- Only control panels selected in the [[camera icon]Control Settings] menu will be displayed.

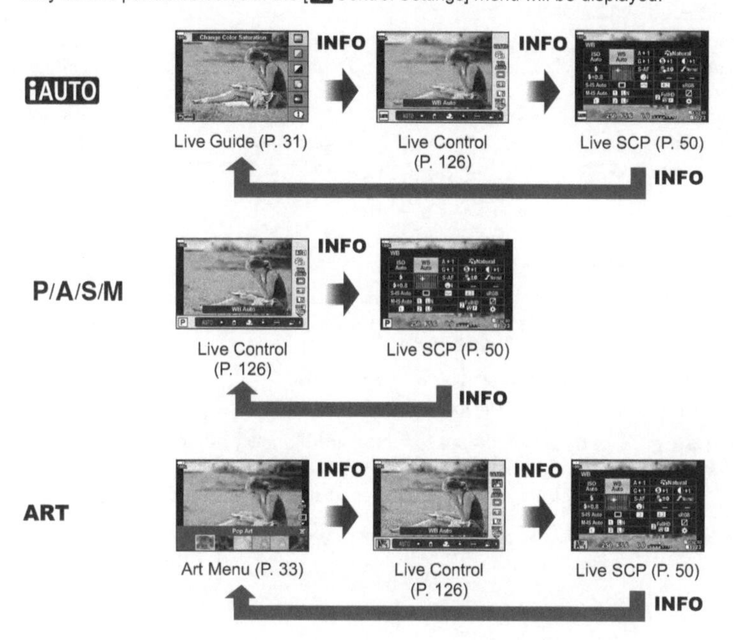

- For movie mode, set with [[movie icon] Control Settings] of [[movie icon] Display Settings] (P. 101).

Live control

Available settings

* Available in movie mode

- Some functions cannot be used depending on the shooting mode.
- When the controls in [Control Settings] is set to [Live Control], you can use live control even in iAUTO, **P**, **A**, **S**, **M**, **ART** modes (P. 115).

1 Press the OK button to display the live control.

- Press the OK button again to hide live control.

2 Use △▽ to move the cursor to the desired function, then use ◁▷ to select it and press the OK button.

- The setting is confirmed if you leave the camera as-is for 8 seconds.

Adding information displays

MENU ➜ ✲ ➜ D1 ➜ [▦/Info Settings]

▶ Info (Playback information displays)

Use [▶ Info] to add the following playback information displays. The added displays are displayed by repeatedly pressing the **INFO** button during playback. You can also choose to not show displays that appear at the default setting.

Histogram display

Highlight & Shadow display

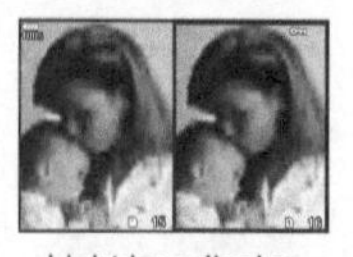

Light box display

Highlight & Shadow display

Areas above the upper limit of brightness for the image are shown in red, those below the lower limit in blue. ☞ [Histogram Settings] (P. 116)

Light box display

Compare two images side-by-side. Press the Ⓞ button to select the image on the opposite side of the display.

- The base image is displayed on the right. Use the front dial to select an image and press the Ⓞ to move the image to the left. The image to be compared to the image on the left can be selected on the right. To choose a different base image, highlight the right frame and press the Ⓞ.
- To change the zoom ratio, turn the rear dial. Press the **Fn1** button and then △▽◁▷ to scroll the zoomed-in area, and rotate the front dial to select between images.

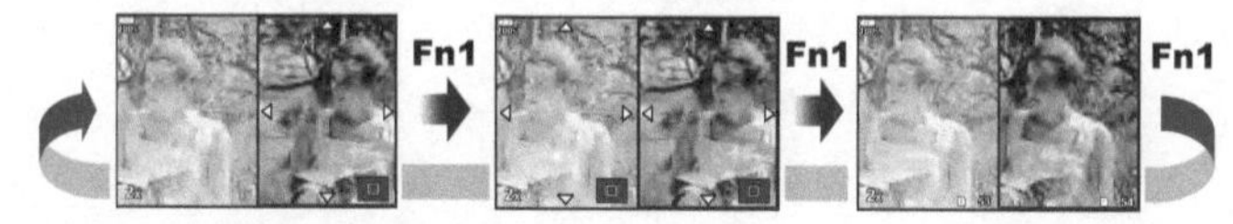

▶Q Info (Magnified playback information display)

The magnified playback information displays can be set with [▶Q Info]. If [Q] (Magnify) is assigned to a button with Button Function (P. 66) in advance, the set displays can be switched by repeatedly pressing the Q button during playback. You can also choose to not show displays that appear at the default setting.

4 Menu functions (Custom menus)

LV-Info (Shooting information displays)

You can add the Highlight&Shadow display screen to [LV-Info]. The added displays are displayed by repeatedly pressing the **INFO** button during shooting. You can also choose to not show displays that appear at the default setting.

▦ Settings (Index/calendar display)

You can change the number of frames to be displayed on the index display and set to not display the screens that are set to be displayed by default with [▦ Settings]. Screens with a check can be selected on the playback screen using the rear dial.

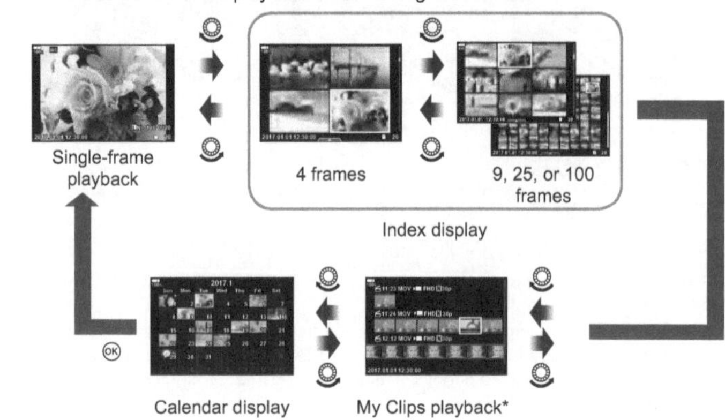

* If one or more My Clips have been created, it will be displayed here (P. 72).

4 Menu functions (Custom menus)

Shooting self-portraits using the selfie assist menu

MENU ➜ ✲ ➜ D3 ➜ [Selfie Assist]

When the monitor is in the selfie position, you can display a convenient touch menu.

1 Select [On] for [Selfie Assist] in Custom Menu D3.

2 Turn the monitor towards you.

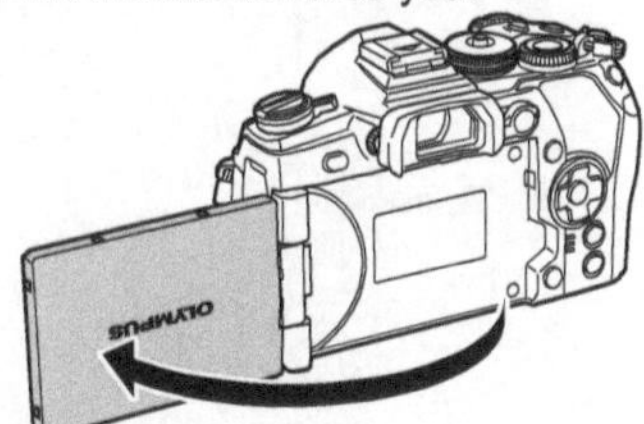

Self portrait menu

- The self portrait menu is displayed in the monitor.

[icon: e-Portrait Off]	**One touch e-Portrait**	Turning this on makes skin look smooth and translucent. Valid only during iAUTO mode (iAUTO).
[icon: shutter]	**Touch shutter**	When the icon is tapped, the shutter is released about 1 second later.
[icon: self-timer Off]	**One touch custom self-timer**	Shoot 3 frames using the self-timer. You can set the number of times the shutter is released and the interval between each release using [Custom Self-timer] (P. 46, 54).

3 Frame the shot.

- Be careful that your fingers or the camera strap do not obstruct the lens.

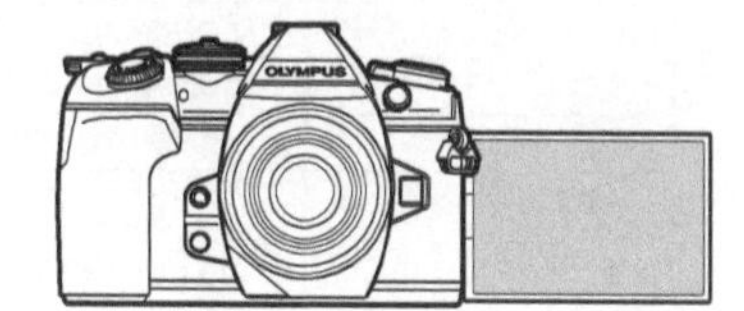

4 Tap [icon: shutter] and shoot.

- The shot image will be displayed on the monitor.
- You can also shoot by tapping the subject displayed in the monitor, or by pressing the shutter button.

Viewing camera images on TV

MENU → ✲ → D4 → [HDMI]

Use the separately sold cable with the camera to playback recorded images on your TV. This function is available during shooting. Connect the camera to an HD TV using an HDMI cable to view high-quality images on a TV screen.

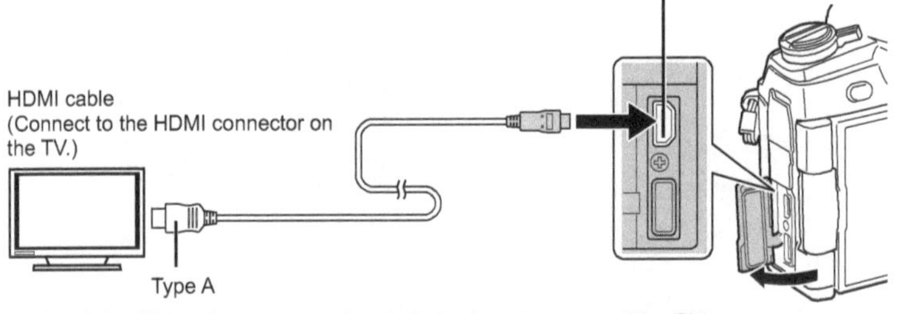

Connect the TV and camera and switch the input source of the TV.

- When an HDMI cable is connected, images are displayed on both the television and the camera monitor. Information is displayed on the television screen only.
- For details on changing the input source of the TV, refer to the TV's instruction manual.
- Depending on the TV's settings, the displayed images and information may become cropped.
- If the camera is connected via an HDMI cable, you will be able to choose the digital video signal type. Choose a format that matches the input format selected with the TV.

C4K	C4K via HDMI output.
4K	Priority is given to 4K HDMI output.
1080p	Priority is given to 1080p HDMI output.
720p	Priority is given to 720p HDMI output.
480p/576p	480p/576p HDMI output.

- Do not connect the camera to other HDMI output devices. Doing so may damage the camera.
- HDMI output is not performed while connected via USB to a computer or printer.
- When [Output Mode] is set to [Record Mode] (P. 101), movies are output in the record mode set for movie. The images cannot be displayed on the connected TV if the TV does not support the record mode.
- 1080p output will be used in place of [4K] or [C4K] while the camera is being used to take photographs.

Using the TV remote control

The camera can be operated by a TV remote control when connected to a TV that supports HDMI control. ☞ [HDMI] (P. 117)

The camera monitor turns off.

- You can operate the camera by following the operation guide displayed on the TV.
- During single-frame playback, you can display or hide the information display by pressing the **"Red"** button, and display or hide the index display by pressing the **"Green"** button.
- Some televisions may not support all features.

Shutter speeds when the flash fires

MENU ➜ ✲ ➜ F ➜ [⚡ X-Sync.]/[⚡ Slow Limit]

You can set shutter speed conditions for when the flash fires.

Shooting mode	Flash firing shutter speed	Upper limit	Lower limit
P	The camera automatically sets the shutter speed.	[⚡ X-Sync.] setting	[⚡ Slow Limit] setting*
A			
S	The set shutter speed		No lower limit
M			

* Extends up to 60 seconds when slow synchronization is set.

Combinations of JPEG image sizes and compression rates

MENU ➜ ✲ ➜ G ➜ [◀:· Set]

You can set the JPEG image quality by combining image size and compression rate.

Image size		Compression rate				Application
Name	Pixel Count	SF (Super Fine)	F (Fine)	N (Normal)	B (Basic)	
L (Large)	5184×3888*	LSF	LF*	LN*	LB	Select for the print size
M (Middle)	3200×2400*	MSF	MF	MN*	MB	
	2560×1920					
	1920×1440					
	1600×1200					
S (Small)	1280×960*	SSF	SF	SN*	SB	For small prints and use on a website
	1024×768					

* Default

4

Menu functions (Custom menus)

Setting the card to record to

MENU ➔ ✲ ➔ H1 ➔ [Card Slot Settings]

When there are cards in both slots 1 and 2, you can select which card to record still images and movies to.

1 Select [Card Slot Settings] in Custom Menu H1 and press the ⊛ button.

2 Use △▽ to select the item and press ▷.

- Use △▽ to select the setting and press the ⊛ button.

Save Settings	Sets the recording method for still images. ☞ "Setting the saving method for shooting data (Save Settings)" (P. 54)
Save Slot	Sets the card for recording still images. This is operative when [Save Settings] is set to [Standard] or [Auto Switch].
Save Slot (movie)	Sets the destination for recording movies.
▶ Slot	Selects the card for still image playback when [Save Settings] is set to [Dual Independent ↓□], [Dual Independent ↑□], [Dual Same ↓□], or [Dual Same ↑□].
Assign Save Folder	Sets the card save destination folder.

Assign Save Folder

1 Select [Assign Save Folder] and press ▷.

2 Select [Assign] and press ▷.

3 Select a folder and press the ⊛ button.

- If you select [New Folder], specify the 3-digit folder number and press the ⊛ button.
- If you select [Existing Folder], use △▽ to select the existing folder and press the ⊛ button.
The first 2 frames and the last frame in the selected folder are displayed.

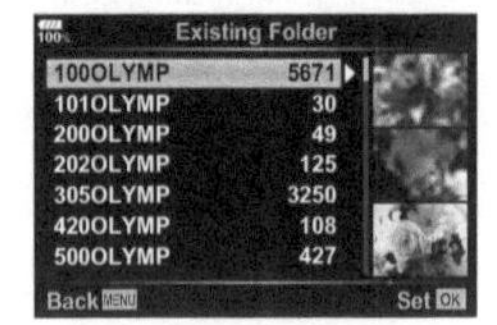

Saving lens info

MENU ➔ ✲ ➔ H1 ➔ [Lens Info Settings]

Store lens info for up to 10 lenses that do not supply info automatically to the camera.

1 Select [Create Lens Information] for [Lens Info Settings] in Custom Menu H1.

2 Select [Lens Name] and enter a lens name. After entering a name, highlight [END] and press the ⊛ button.

3 Use △▽◁▷ to choose the [Focal Length].

4 Use △▽◁▷ to choose the [Aperture Value].

5 Select [Set] and press the ⊛ button.

- The lens will be added to the lens info menu.
- When a lens that does not supply info automatically is attached, the info used is indicated by ✔. Highlight a lens with a ✔ icon and press the ⊛ button.

Selecting the display style of the viewfinder

MENU ➜ ✲ ➜ ◧ ➜ [EVF Style]

Style 1/2: Displays only main items such as shutter speed and aperture value
Style 3: Displays same as the monitor

Style 1/ Style 2

Style 3

■ Viewfinder display when shooting using the viewfinder (Style 1/Style 2)

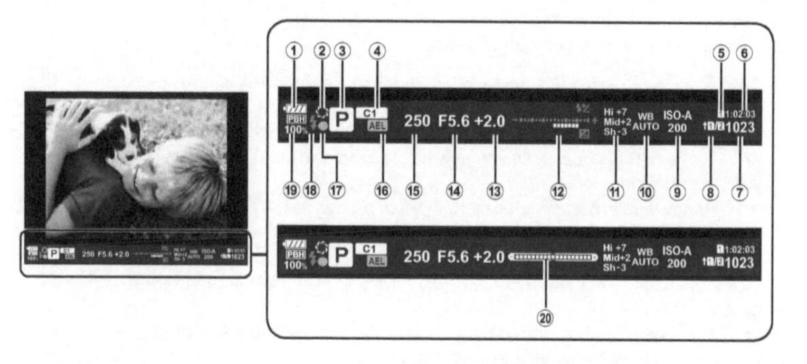

① Battery check
- : Ready for use.
- : Low battery
- : Battery is running low.
- : Charging required.

② ....P. 67
③ Shooting mode..........................P. 24–37
④ Assign to Custom Mode............P. 35, 87
⑤ Save Slot.................................P. 132
⑥ Available recording time ....P. 148
⑦ Number of storable still pictures ...P. 148
⑧ Save Settings ....P. 54
⑨ ISO sensitivity...........................P. 42, 51
⑩ White balance...........................P. 42, 52
⑪ Highlight&Shadow Control ....P. 66
⑫ Top: Flash intensity control ....P. 60
Bottom: Exposure compensation indicator...P. 39
⑬ Exposure compensation value........P. 39
⑭ Aperture value .... P. 26–29
⑮ Shutter speed ....P. 26–29
⑯ AE Lock AEL ....P. 45, 123
⑰ AF confirmation mark......................P. 25
⑱ Flash (blinks: charging in progress) ....P. 57
⑲ PBH (displayed when using the power battery holder power).....................P. 150
⑳ Level gauge (displayed by pressing the shutter button down halfway)

You can change the viewfinder display style. ☞ [EVF Style] (P. 121)

5 Connecting the camera to a smartphone

By connecting to a smartphone through this camera's wireless LAN function and using the specified app, you can enjoy even more features during and after shooting.

Things you can do with the specified app, OLYMPUS Image Share (OI.Share)

- Camera image transfer to a smartphone
  You can load images in the camera to a smartphone.
- Remote shooting from a smartphone
  You can remotely operate the camera and shoot using a smartphone.
- Beautiful image processing
  You can apply art filters and add stamps on images loaded to a smartphone.
- Addition of GPS tags to camera images
  You can add GPS tags to images simply by transferring the GPS log saved on the smartphone to the camera.

For details, visit the address below:
http://app.olympus-imaging.com/oishare/

- Before using the wireless LAN function, read "Using the wireless LAN function" (P. 183).
- If using the wireless LAN function in a country outside the region where the camera was purchased, there is a risk that the camera will not conform to the wireless communication regulations of that country. Olympus will not be held responsible for any failure to meet such regulations.
- As with any wireless communication, there is always a risk of interception by a third party.
- The wireless LAN function on the camera cannot be used to connect to a home or public access point.
- The transmitting antenna is located inside the grip. Keep the antenna away from metal objects whenever possible.
- During wireless LAN connection, the battery will run down faster. If the battery is running low, the connection may be lost during a transfer.
- Connection may be difficult or slow in proximity to devices that generate magnetic fields, static electricity or radio waves, such as near microwaves, cordless telephone.

- While connected to a smartphone running OI.Share, the camera will function as if [Standard] is selected for [Card Slot Settings] > [Save Slot] (P. 132), and OI.Share will have access only to the card in the slot currently selected for [Save Slot]. The slot can not be changed using OI.Share.
- If only one card is inserted, OI.Share will access it automatically.
- Movies are recorded to the card used for photographs, regardless of the option selected for [Save Slot].

Connecting to a smartphone

Connect to a smartphone. Start the OI.Share App installed on your smartphone.

1 Select [Connection to Smartphone] in the ▶ Playback Menu and press the Ⓞ button.

- You can also connect by tapping Wi-Fi in the monitor.

2 Following the guide displayed on the monitor, proceed with the Wi-Fi settings.

- The SSID, password and QR code are displayed on the monitor.

3 Start OI.Share on your smartphone, and read the QR code displayed on the camera monitor.

- Connection will be performed automatically.
- Some smartphones will need to be configured manually after reading the QR code.
- If you are unable to read the QR code, enter the SSID and password in the Wi-Fi settings of your smartphone to connect. For how to access the Wi-Fi settings on your smartphone, please see your smartphone operating instructions.

4 To end the connection, press **MENU** on the camera or tap [End Wi-Fi] on the monitor screen.

- You can also end the connection with OI.Share or by turning off your camera.
- The connection ends.

Transferring images to a smartphone

You can select images in the camera and load them to a smartphone. You can also use the camera to select images you want to share in advance. ☞ "Setting a transfer order on images (Share Order)" (P. 82)

1 Connect the camera to a smartphone (P. 135).
- You can also connect by tapping Wi-Fi in the monitor.

2 Launch OI.Share and tap the Image Transfer button.
- The images in the camera are displayed in a list.

3 Select the pictures you want to transfer and tap the Save button.
- When saving is completed, you can turn off the camera from the smartphone.

Shooting remotely with a smartphone

You can shoot remotely by operating the camera with a smartphone.
This is available only in [Private].

1 Start [Connection to Smartphone] on the camera.
- You can also connect by tapping Wi-Fi in the monitor.

2 Launch OI.Share and tap the Remote button.

3 Tap the shutter button to shoot.
- The image taken is saved on the memory card in the camera.

- Available shooting options are partially limited.

Adding position information to images

You can add GPS tags to images that were taken while the GPS log was saving by transferring the GPS log saved on the smartphone to the camera.
This is available only in [Private].

1 Before beginning to shoot, launch OI.Share and turn on the switch on the Add Location button to begin saving the GPS log.

- Before beginning to save the GPS log, the camera must be connected to OI.Share once to synchronize the time.
- You can use the phone or other apps while the GPS log is saving. Do not terminate OI.Share.

2 When shooting is complete, turn off the switch on the Add Location button. Saving the GPS log is complete.

3 Start [Connection to Smartphone] on the camera.

- You can also connect by tapping Wi-Fi in the monitor.

4 Transfer the saved GPS log to the camera using OI.Share.

- GPS tags are added to the images in the memory card based on the transferred GPS log.
- 🛰 is displayed on images to which position information has been added.

- Addition of location information can only be used with smartphones that have a GPS function.
- Position information cannot be added to movies.

Changing the connection method

There are two ways to connect to a smartphone. With [Private] the same settings are used to connect every time. With [One-Time] different settings are used each time. You may find it convenient to use [Private] when connecting to your own smartphone and [One-Time] when transferring images to a friend's smartphone etc.
The default setting is [Private].

1 Select [Wi-Fi Settings] in the γ Setup Menu and press the Ⓞ button.

2 Select [Wi-Fi Connect Settings] and press ▷.

3 Select the wireless LAN connection method and press the Ⓞ button.

- [Private]: Connect to one smartphone (connects automatically using the settings after the initial connection). All OI.Share functions are available.
- [One-Time]: Connect to multiple smartphones (connects using different connection settings each time). Only OI.Share's image transfer function is available. You can view only images that are set for share order using the camera.
- [Select]: Select which method to use each time.
- [Off]: The Wi-Fi function is turned off.

Changing the password

Change the password used for [Private].

1 Select [Wi-Fi Settings] in the Setup Menu and press the OK button.
2 Select [Private Password] and press ▷.
3 Follow the operation guide and press the ◉ button.
   - A new password will be set.

Cancelling a share order

Cancel share orders that are set on images.

1 Select [Wi-Fi Settings] in the Setup Menu and press the OK button.
2 Select [Reset share Order] and press ▷.
3 Select [Yes] and press the OK button.
   - The share order for images saved on the card being used for playback will be canceled.

Initializing wireless LAN settings

Initializes content of [Wi-Fi Connect Settings].

1 Select [Wi-Fi Settings] in the Setup Menu and press the OK button.
2 Select [Reset Wi-Fi Settings] and press ▷.
3 Select [Yes] and press the OK button.

6 Connecting the camera to a computer and a printer

Connecting the camera to a computer

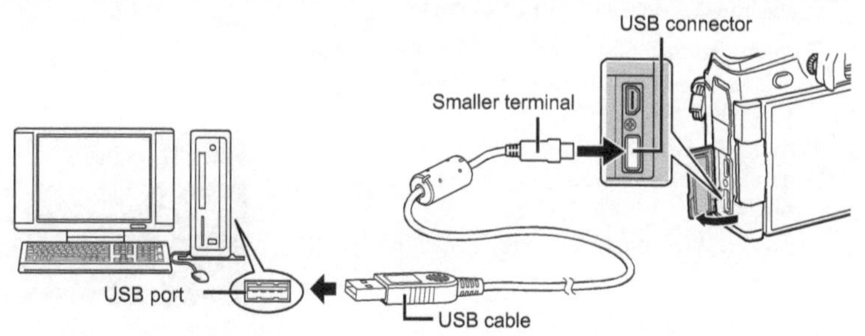

- If nothing is displayed on the camera screen even after connecting the camera to the computer, the battery may be exhausted. Use a full-charged battery.
- When the camera is turned on, a dialog should be displayed in the monitor prompting you to choose a host. If it is not, select [Auto] for [USB Mode] (P. 117) in the camera custom menus.

Copying pictures to a computer

The following operating systems are compatible with the USB connection:

Windows: **Windows Vista SP2/Windows 7 SP1/Windows 8/ Windows 8.1/Windows 10**

Macintosh: **Mac OS X v10.8 - v10.11**

1 Turn the camera off and connect it to the computer.

- The location of the USB port varies with the computer. For details, refer to your computer's manual.

2 Turn on the camera.

- The selection screen for the USB connection is displayed.

3 Press △▽ to select [Storage]. Press the Ⓞ button.

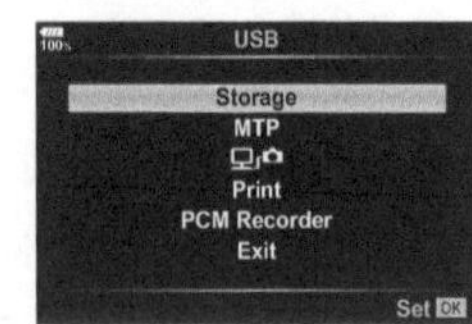

4 The computer recognizes the camera as a new device.

- If you are using Windows Photo Gallery, choose [MTP] in Step 3.
- Data transfer is not guaranteed in the following environments, even if your computer is equipped with a USB port.
  Computers with a USB port added by means of an extension card, etc., computers without a factory-installed OS, or home-built computers
- Camera controls cannot be used while the camera is connected to a computer.
- When [🖵/📷] is selected, camera controls can be used even while the camera is connected to a computer.
- If the dialog shown in Step 2 is not displayed when the camera is connected to a computer, select [Auto] for [USB Mode] (P. 117) in the camera custom menus.

6 Connecting the camera to a computer and a printer

Installing the PC software

The photographs and movies you have shot with your camera can be transferred to a computer and viewed, edited, and organized using the OLYMPUS Viewer 3 offered by OLYMPUS.

- To install OLYMPUS Viewer 3, download it from http://support.olympus-imaging.com/ov3download/ and follow the on-screen instructions.
- Visit the website above for system requirements and installation instructions.
- You will be required to enter the product serial number prior to download.

Installing the OLYMPUS Digital Camera Updater

Camera firmware updates can only be performed using the OLYMPUS Digital Camera Updater. Download the updater from the website below and install it according to the on-screen instructions.

http://oup.olympus-imaging.com/ou1download/index/

User Registration

Visit the OLYMPUS website for information on registering your OLYMPUS products.

Direct printing (PictBridge)

By connecting the camera to a PictBridge-compatible printer with the USB cable, you can print out recorded pictures directly.

1 Connect the camera to the printer using the supplied USB cable and turn the camera on.

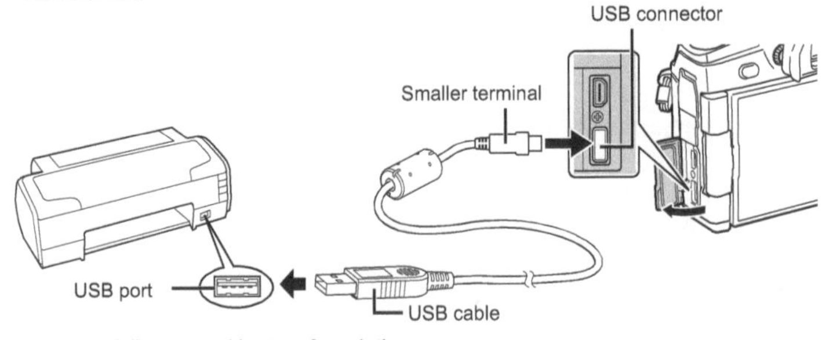

- Use a fully charged battery for printing.
- When the camera is turned on, a dialog should be displayed in the monitor prompting you to choose a host. If it is not, select [Auto] for [USB Mode] (P. 117) in the camera custom menus.

2 Use △▽ to select [Print].

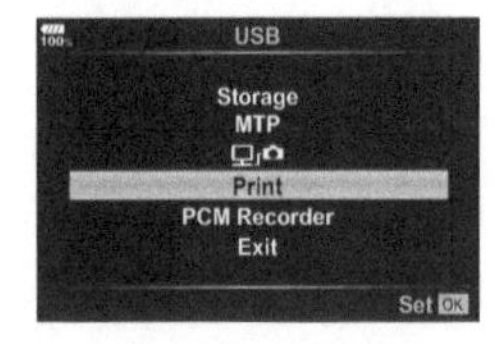

- [One Moment] will be displayed, followed by a print-mode selection dialog.
- If the screen is not displayed after a few minutes, disconnect the USB cable and start again from Step 1.

Proceed to "Custom printing" (P. 143)

- RAW images and movies cannot be printed.

Easy printing

Use the camera to display the picture you wish to print before connecting the printer via the USB cable.

1 Use ◁▷ to display the pictures you want to print on the camera.

2 Press ▷.

- The picture selection screen appears when printing is completed. To print another picture, use ◁▷ to select the image and press the OK button.
- To exit, unplug the USB cable from the camera while the picture selection screen is displayed.

Custom printing

1 Connect the camera to the printer using the supplied USB cable and turn the camera on.

- When the camera is turned on, a dialog should be displayed in the monitor prompting you to choose a host. If it is not, select [Auto] for [USB Mode] (P. 117) in the camera custom menus.

2 Follow the operation guide to set a print option.

Selecting the print mode

Select the type of printing (print mode). The available print modes are as shown below.

Print	Prints selected pictures.
All Print	Prints all the pictures stored in the card and makes one print for each picture.
Multi Print	Prints multiple copies of one image in separate frames on a single sheet.
All Index	Prints an index of all the pictures stored in the card.
Print Order	Prints according to the print reservation you made. If there is no picture with print reservation, this is not available.

Setting the print paper items

This setting varies with the type of printer. If only the printer's STANDARD setting is available, you cannot change the setting.

Size	Sets the paper size that the printer supports.
Borderless	Selects whether the picture is printed on the entire page or inside a blank frame.
Pics/Sheet	Selects the number of pictures per sheet. Displayed when you have selected [Multi Print].

Selecting pictures you want to print

Select pictures you want to print. The selected pictures can be printed later (single-frame reservation) or the picture you are displaying can be printed right away.

Print OK	Prints the currently displayed picture. If there is a picture that [Single Print ▲] reservation has already been applied to, only that reserved picture will be printed.
Single Print ▲	Applies print reservation to the currently displayed picture. If you want to apply reservation to other pictures after applying [Single Print ▲], use ◁▷ to select them.
More ▼	Sets the number of prints and other items for the currently displayed picture, and whether or not to print it. For operation, refer to "Setting printing data" in the next section.

Setting printing data

Select whether to print printing data such as the date and time or file name on the picture when printing. When the print mode is set to [All Print], select [Option Set].

🖶×	Sets the number of prints.
Date	Prints the date and time recorded on the picture.
File Name	Prints the file name recorded on the picture.
⌗	Trims the picture for printing. Use the front dial (🎛) to choose the crop size and △▽◁▷ to specify the crop position.

3 Once you have set the pictures for printing and printing data, select [Print], then press the Ⓞ button.

- The setting will be applied to the images saved on the card being used for playback.
- To stop and cancel printing, press the Ⓞ button. To resume printing, select [Continue].

■ Cancelling printing

To cancel printing, highlight [Cancel] and press the Ⓞ button. Note that any changes to the print order will be lost; to cancel printing and return to the previous step, where you can make changes to the current print order, press **MENU**.

Print order (DPOF)

You can save digital "print orders" to the memory card listing the pictures to be printed and the number of copies of each print. You can then have the pictures printed at a print shop that supports DPOF or print the pictures yourself by connecting the camera directly to a DPOF printer. A memory card is required when creating a print order.

Creating a print order

1 Press the Ⓞ button during playback and select [Print Order].

2 Select [🖶] or [🖶ALL] and press the Ⓞ button.

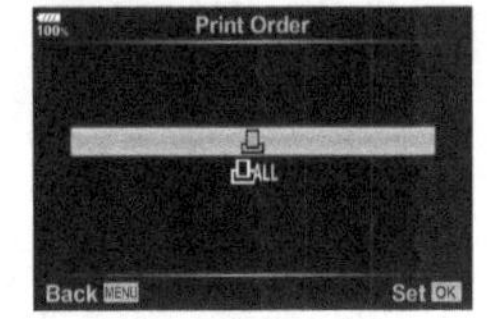

Individual picture

Press ◁▷ to select the frame that you want to set as print reservation, then press △▽ to set the number of prints.

- To set print reservation for several pictures, repeat this step. Press the Ⓞ button when all the desired pictures have been selected.

All pictures

Select [🖶ALL] and press the Ⓞ button.

3 Select the date and time format and press the ⓞ button.

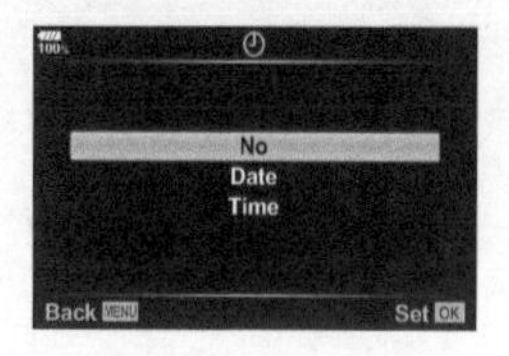

No	The pictures are printed without the date and time.
Date	The pictures are printed with the shooting date.
Time	The pictures are printed with the shooting time.

- When printing images, the setting cannot be changed between the images.

4 Select [Set] and press the ⓞ button.

- The setting will be applied to the images saved on the card being used for playback.

- The camera cannot be used to modify print orders created with other devices. Creating a new print order deletes any existing print orders created with other devices.
- Print orders cannot include RAW images or movies.

Removing all or selected pictures from the print order

You can reset all print reservation data or just the data for selected pictures.

1 Press the ⓞ button during playback and select [Print Order].

2 Select [凸] and press the ⓞ button.

- To remove all pictures from the print order, select [Reset] and press the ⓞ button. To exit without removing all pictures, select [Keep] and press the ⓞ button.

3 Press ◁▷ to select images you wish to remove from the print order.

- Use ▽ to set the number of prints to 0. Press the ⓞ button once you have removed all the desired pictures from the print order.

4 Select the date and time format and press the ⓞ button.

- This setting is applied to all frames with print reservation data.
- The setting will be applied to the images saved on the card being used for playback.

5 Select [Set] and press the ⓞ button.

7 Cautions

Battery and charger

- The camera uses a single Olympus lithium-ion battery. Never use any battery other than genuine OLYMPUS batteries.
- The camera's power consumption varies widely with usage and other conditions.
- As the following consume a lot of power even without shooting, the battery will be drained quickly.
  - Performing auto focus repeatedly by pressing the shutter button halfway in shooting mode.
  - Displaying images on the monitor for a prolonged period.
  - When connected to a computer or printer.
- When using a drained battery, the camera may turn off without the low battery warning being displayed.
- The battery will not be fully charged at the time of purchase. Charge the battery using the provided charger before use.
- If you will not be using the camera for a prolonged period (one month or more), remove the battery from the camera.
  Make sure not to leave the battery inside the camera for a long time, or else the battery life may be shortened or the battery may become unusable.
- The normal charging time using the provided charger is approximately 2 hours (estimated).
- Do not attempt to use chargers not specifically designated for use with the supplied battery, or to use batteries not specifically designated for use with the supplied charger.
- There is a risk of explosion if the battery is replaced with the incorrect battery type.
- Dispose of the used battery following the instructions "CAUTION" (P. 182) in the instruction manual.

Using the optional AC adapter

The optional AC-5 AC adapter can be used with the power battery holder (HLD-9). (P. 150)
Use only the appropriate AC adapter. Do not use the power cord included with the AC adapter with any other products.

Using your charger abroad

- The charger can be used in most home electrical sources within the range of 100 V to 240 V AC (50/60 Hz) around the world. However, depending on the country or area you are in, the AC wall outlet may be shaped differently and the charger may require a plug adapter to match the wall outlet.
- Do not use commercially available travel adapters as the charger may malfunction.

Usable cards

In this manual, all storage devices are referred to as "cards." The following types of SD memory card (commercially available) can be used with this camera: SD, SDHC, SDXC, and Eye-Fi. For the latest information, please visit the Olympus website.

SD card write protect switch

The SD card body has a write protect switch. Setting the switch to "LOCK" prevents data being written to the card. Return the switch to the unlock position enable writing.

- The data in the card will not be erased completely even after formatting the card or deleting the data. When discarding, destroy the card to prevent leakage of personal information.
- Use the Eye-Fi card in compliance with the laws and regulations of the country where the camera is used. Remove the Eye-Fi card from the camera or disable the card functions in airplanes and other locations where use is prohibited. ☞ [Eye-Fi] (P. 122)
- The Eye-Fi card may become hot during use.
- When using an Eye-Fi card, the battery may run out faster.
- When using an Eye-Fi card, the camera may function more slowly.
- Failure may occur during My Clips shooting. Please turn off the card function in this case.
- Setting the write protect switch to "LOCK" may limit such functions as clip shooting and playback.

Record mode and file size/number of storable still pictures

The file size in the table is approximate for files with a 4:3 aspect ratio.

Record mode	Image size (Pixel Count)	Compression	File format	File size (MB)	Number of storable still pictures*
50M F+RAW	10368×7776	Uncompressed	ORF	181.5	42
	8160×6120	1/4	JPEG		
	5184×3888	Uncompressed	ORI		
25M F+RAW	10368×7776	Uncompressed	ORF	169.5	44
	5760×4320	1/4	JPEG		
	5184×3888	Uncompressed	ORI		
50M F	8160×6120	1/4	JPEG	Approx. 21.7	317
25M F	5760×4320	1/4	JPEG	Approx. 10.9	630
RAW	5184×3888	Loss-less compression	ORF	Approx. 21.5	341
L SF		1/2.7	JPEG	Approx. 13.1	527
L F		1/4		Approx. 8.9	774
L N		1/8		Approx. 4.6	1506
L B		1/12		Approx. 3.1	2219
M SF	3200×2400	1/2.7		Approx. 5.1	1348
M F		1/4		Approx. 3.6	1952
M N		1/8		Approx. 1.9	3698
M B		1/12		Approx. 1.4	5194
M SF	2560×1920	1/2.7		Approx. 3.4	2051
M F		1/4		Approx. 2.4	2941
M N		1/8		Approx. 1.3	5424
M B		1/12		Approx. 1.0	7397
M SF	1920×1440	1/2.7		Approx. 2.0	3487
M F		1/4		Approx. 1.4	4882
M N		1/8		Approx. 0.9	8418
M B		1/12		Approx. 0.7	11096
M SF	1600×1200	1/2.7		Approx. 1.5	4786
M F		1/4		Approx. 1.1	6597
M N		1/8		Approx. 0.7	11096
M B		1/12		Approx. 0.5	13562
S SF	1280×960	1/2.7		Approx. 1.0	6781
S F		1/4		Approx. 0.8	9041
S N		1/8		Approx. 0.5	14360
S B		1/12		Approx. 0.4	17437
S SF	1024×768	1/2.7		Approx. 0.8	9389
S F		1/4		Approx. 0.6	12206
S N		1/8		Approx. 0.3	30515
S B		1/12		Approx. 0.2	40687

* Assumes a 8GB SD card.

- The number of storable still pictures may change according to the subject, whether or not print reservations have been made, and other factors. In certain instances, the number of storable still pictures displayed on the monitor will not change even when you take pictures or erase stored images.
- The actual file size varies according to the subject.
- The maximum number of storable still pictures displayed on the monitor is 9999.
- For the available recording time for movies, see the Olympus website.

Interchangeable lenses

Choose a lens according to the scene and your creative intent. Use lenses designed exclusively for the Micro Four Thirds system and bearing the M.ZUIKO DIGITAL label or the symbol shown at right. With an adapter, you can also use Four Thirds System and OM System lenses. The optional adapter is required.

- When you attach or remove the body cap and lens from the camera, keep the lens mount on the camera pointed downward. This helps prevent dust and other foreign matter from getting inside the camera.
- Do not remove the body cap or attach the lens in dusty places.
- Do not point the lens attached to the camera toward the sun. This may cause the camera to malfunction or even ignite due to the magnifying effect of sunlight focusing through the lens.
- Be careful not to lose the body cap and rear cap.
- Attach the body cap to the camera to prevent dust from getting inside when no lens is attached.

■ Lens and camera combinations

<table>
<tr><th>Lens</th><th>Camera</th><th>Attachment</th><th>AF</th><th>Metering</th></tr>
<tr><td>Micro Four Thirds system lens</td><td rowspan="3">Micro Four Thirds system camera</td><td>Yes</td><td>Yes</td><td>Yes</td></tr>
<tr><td>Four Thirds system lens</td><td rowspan="2">Attachment possible with mount adapter</td><td>Yes*1</td><td>Yes</td></tr>
<tr><td>OM System lenses</td><td>No</td><td>Yes*2</td></tr>
<tr><td>Micro Four Thirds system lens</td><td>Four Thirds System Camera</td><td>No</td><td>No</td><td>No</td></tr>
</table>

*1 AF is not operational when recording movies.
*2 Accurate metering is not possible.

HLD-9 Power Battery Holder

This can be used along with the battery in the camera body to extend the camera's operating time. You can assign functions to the dial and **B-Fn** button in the Custom Menu. The optional AC adapter can be used with the HLD-9.
Make sure to turn the camera off when attaching and removing the holder.

■ Part names

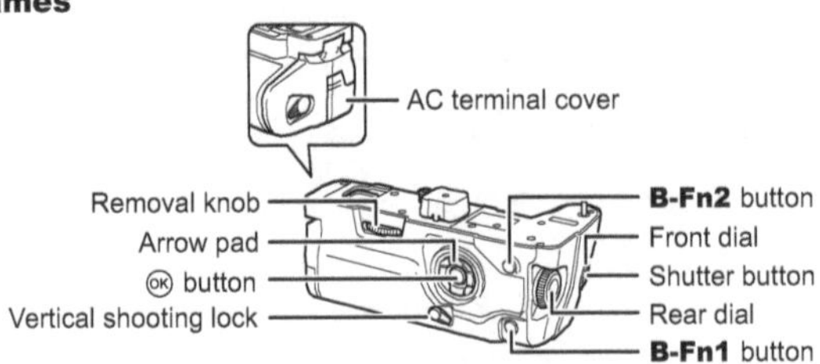

■ Attaching the holder

Remove the PBH cover (a) on the bottom of the camera before attaching the HLD-9. Once attached, make sure the HLD-9 removal knob is tightly secured. Make sure to attach the PHB cover to the camera when not using the HLD-9.

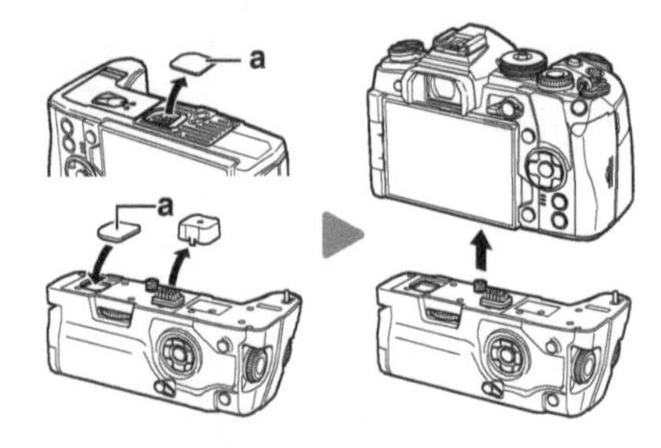

■ Loading the battery

Use BLH-1 battery. Once you load the battery, make sure to lock the battery cover.

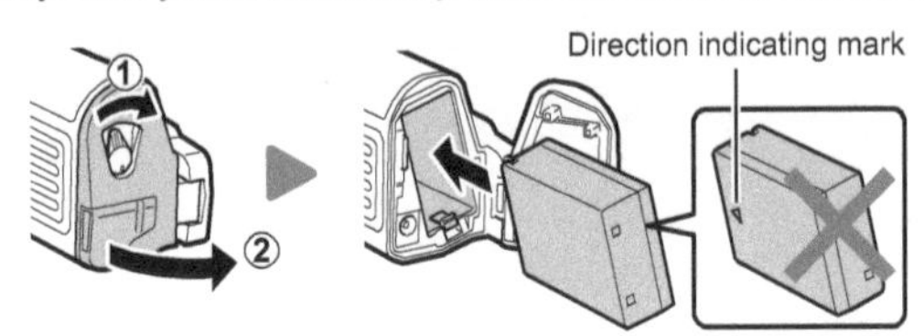

■ Using the AC adapter

Open the AC terminal cover and plug the AC adapter into the AC terminal.

■ Using the dials and buttons

You can set the HLD-9 dial and **B-Fn** button functions in [📷 Button Function] and [🎥 Button Function] of the Custom Menu. ☞ "Assigning functions to buttons (Button Function)" (P. 66), [📷 Button Function] (P. 113), [🎥 Button Function] (P. 100)

■ Main Specifications (HLD-9)

Power Supply	Battery: BLH-1 Lithium-ion Battery x 1 AC power: AC-5 AC Adapter
Dimensions	Approx. 132.7 mm (W) × 55.8 mm (H) × 66.0 mm (D) (5.2" × 2.2" × 2.6")
Weight	Approx. 255 g (9.0 oz.) (excluding battery and terminal cap)
Splash resistance (when attached to camera)	Type Equivalent to IEC Standard publication 60529 IPX1 (under OLYMPUS test conditions)

⚠ Note

- Use only the designated battery and AC adapter. Failure to do so could result in injury, damage to the product, and fire.
- Do not use your nail to turn the removal knob. Doing so could result in injury.
- Only use the camera within the guaranteed operating temperature range.
- Do not use or store the product in dusty or humid areas.
- Do not touch the electrical contacts.
- Use a dry, soft cloth to clean the terminals. Do not clean the product with a damp cloth, thinner, benzine, or any other organic solvents.

External flash units designated for use with this camera

With this camera, you can use one of the separately sold external flash units to achieve a flash suited to your needs. The external flashes communicate with the camera, allowing you to control the camera's flash modes with various available flash control modes, such as TTL-AUTO and Super FP flash. An external flash unit specified for use with this camera can be mounted on the camera by attaching it to the camera's hot shoe. You can also attach the flash to the flash bracket on the camera using the bracket cable (optional). Refer to the documentation provided with the external flash units as well.

The upper limit of the shutter speed is 1/250 sec. when using a flash.

* The sync speed for silent mode and focus bracketing (P. 94) is 1/50 second. The sync speed at ISO sensitivities of 8000 and above and during ISO bracketing (P. 94) is 1/20 second.

Functions available with external flash units

Optional flash	Flash control mode	GN (Guide number) (ISO100)	RC mode
FL-900R	TTL-AUTO, AUTO, MANUAL, FP TTL AUTO, FP MANUAL, MULTI, RC, SL AUTO, SL MANUAL	GN58 (200 mm*1)	✓
FL-600R	TTL-AUTO, AUTO, MANUAL, FP TTL AUTO, FP MANUAL	GN36 (85 mm*1) GN20 (24 mm*1)	✓
FL-300R	TTL-AUTO, MANUAL	GN20 (28 mm*1)	✓
FL-14	TTL-AUTO, AUTO, MANUAL	GN14 (28 mm*1)	–
STF-8	TTL-AUTO, MANUAL	GN8.5	✓

*1 The focal length of the lens that can be used (Calculated based on 35 mm film camera).

Wireless remote control flash photography

External flash units that are designated for use with this camera and have a remote control mode can be used for wireless flash photography. The camera can separately control each of three groups of remote flash units, and the internal flash. See the instruction manuals provided with the external flash units for details.

1 Set the remote flash units to RC mode and place them as desired.

- Turn the external flash units on, press the MODE button, and select RC mode.
- Select a channel and group for each external flash unit.

2 Select [On] for [⚡ RC Mode] in 📷₂ Shooting Menu 2 (P. 86).

- The LV super control panel switches to RC mode.
- You can choose an LV super control panel display by repeatedly pressing the **INFO** button.
- Select a flash mode (note that red-eye reduction is not available in RC mode).

3 Adjust the settings for each group in the LV super control panel.

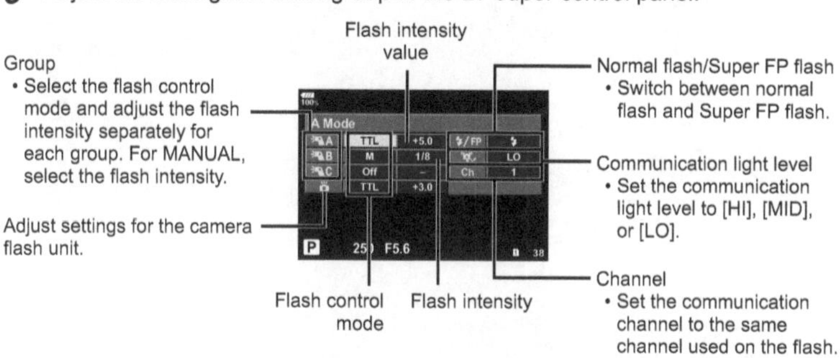

4 Attach the supplied flash unit and turn on the camera.

- After confirming that the built-in and remote flash units have charged, take a test shot.

■ Wireless flash control range

Position the wireless flash units with their remote sensors facing the camera. The following illustration shows the approximate ranges at which the flash units can be positioned. The actual control range varies with local conditions.

- We recommend using a single group of up to three remote flash units.
- Remote flash units cannot be used for second curtain slow synchronization or anti-shock exposures longer than 4 seconds.
- If the subject is too close to the camera, the control flashes emitted by the camera flash may affect exposure (this effect can be reduced by reducing the output of the camera flash by, for example, using a diffuser).
- The upper limit of flash synchronization timing is 1/250 sec. when using the flash in RC mode.

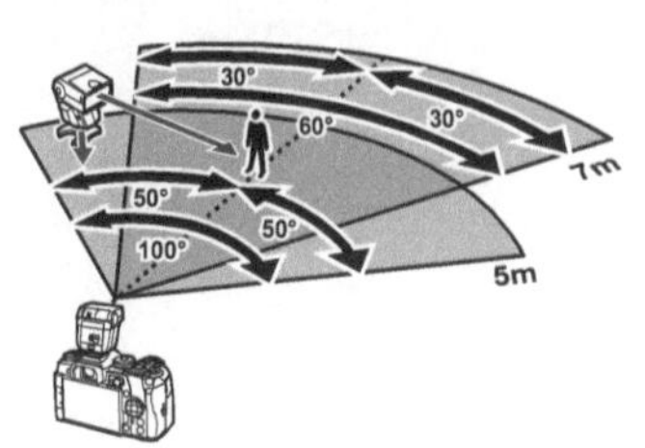

7
Cautions

Other external flash units

Connect a sync cord to the hot shoe or the external flash terminal. When not using the external flash terminal, make sure to attach the cap.

Note the following when using a third-party flash unit mounted on the camera hot shoe:

- Using obsolete flash units that apply currents of more than about 250 V to the X-contact will damage the camera.
- Connecting flash units with signal contacts that do not conform to Olympus specifications may damage the camera.
- Set the Shooting mode to **M**, set the shutter speed to a value no higher than flash synchronous speed, and set ISO sensitivity to a setting other than [AUTO].
- Flash control can only be performed by manually setting the flash to the ISO sensitivity and aperture values selected with the camera. Flash brightness can be adjusted by adjusting either ISO sensitivity or aperture.
- Use a flash with an angle of illumination suited to the lens. Angle of illumination is usually expressed using 35-mm format equivalent focal lengths.

Principal Accessories

Remote cable (RM-CB2)

Use when the slightest camera movement can result in blurred pictures, for example for macro or bulb photography. Connect the cable to the camera remote cable terminal (P. 11).

Converter lenses

Converter lenses attach to the camera lens for quick and easy fish-eye or macro photography. See the OLYMPUS website for information on the lenses that can be used.

Eyecup (EP-13)

You can switch to a large-size eyecup.

Removal

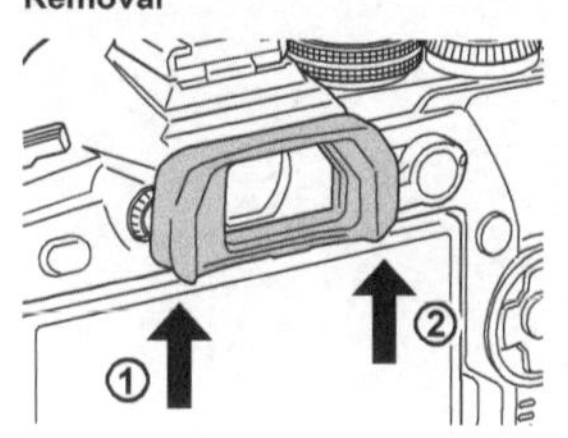

System chart

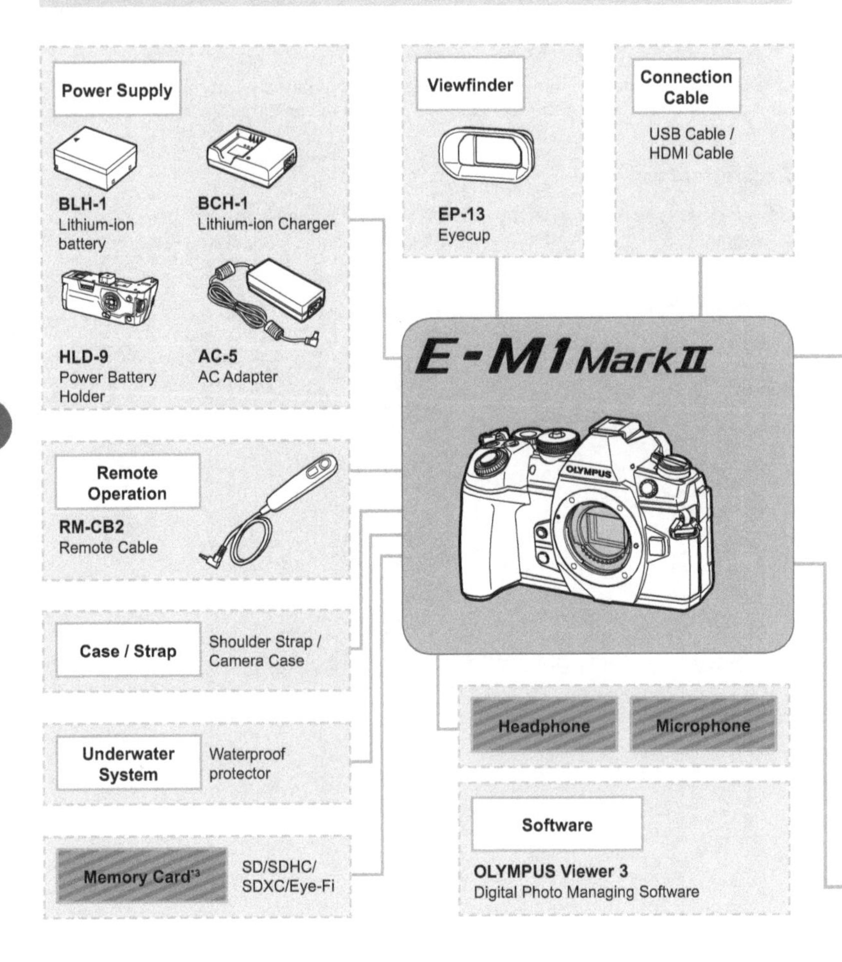

: E-M1 Mark II compatible products

: Commercially available products

For the latest information, please visit the Olympus website.

*1 Not all lenses can be used with adapter. For details, refer to the Olympus Official Web Site. Also, please note that manufacture of OM System Lenses has been discontinued.

*2 For compatible lens, refer to the Olympus Official Web Site.

*3 Use the Eye-Fi card in compliance with the laws and regulations of the country where the camera is used.

*4 Only available for ED 40-150mm f2.8 PRO and ED 300mm f4.0 IS PRO.

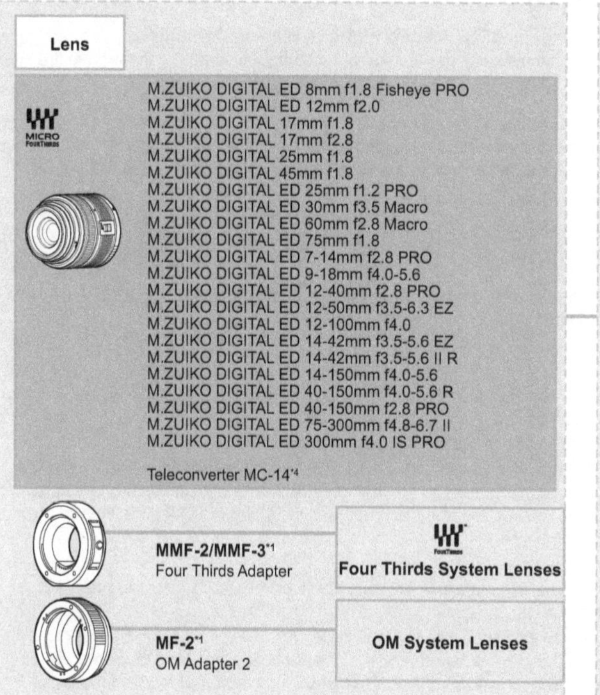
Lens
MICRO FOURTHIRDS
M.ZUIKO DIGITAL ED 8mm f1.8 Fisheye PRO
M.ZUIKO DIGITAL ED 12mm f2.0
M.ZUIKO DIGITAL 17mm f1.8
M.ZUIKO DIGITAL 17mm f2.8
M.ZUIKO DIGITAL 25mm f1.8
M.ZUIKO DIGITAL 45mm f1.8
M.ZUIKO DIGITAL ED 25mm f1.2 PRO
M.ZUIKO DIGITAL ED 30mm f3.5 Macro
M.ZUIKO DIGITAL ED 60mm f2.8 Macro
M.ZUIKO DIGITAL ED 75mm f1.8
M.ZUIKO DIGITAL ED 7-14mm f2.8 PRO
M.ZUIKO DIGITAL ED 9-18mm f4.0-5.6
M.ZUIKO DIGITAL ED 12-40mm f2.8 PRO
M.ZUIKO DIGITAL ED 12-50mm f3.5-6.3 EZ
M.ZUIKO DIGITAL ED 12-100mm f4.0
M.ZUIKO DIGITAL ED 14-42mm f3.5-5.6 EZ
M.ZUIKO DIGITAL ED 14-42mm f3.5-5.6 II R
M.ZUIKO DIGITAL ED 14-150mm f4.0-5.6
M.ZUIKO DIGITAL ED 40-150mm f4.0-5.6 R
M.ZUIKO DIGITAL ED 40-150mm f2.8 PRO
M.ZUIKO DIGITAL ED 75-300mm f4.8-6.7 II
M.ZUIKO DIGITAL ED 300mm f4.0 IS PRO
Teleconverter MC-14*4
MMF-2/MMF-3*1
Four Thirds Adapter
Four Thirds System Lenses
MF-2*1
OM Adapter 2
OM System Lenses

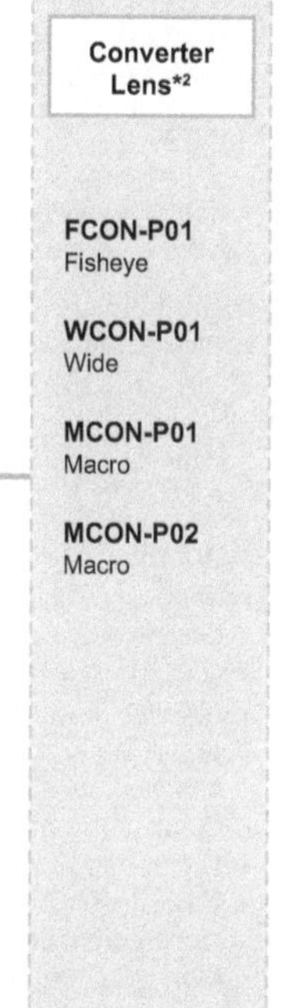
Converter Lens*2
FCON-P01
Fisheye
WCON-P01
Wide
MCON-P01
Macro
MCON-P02
Macro

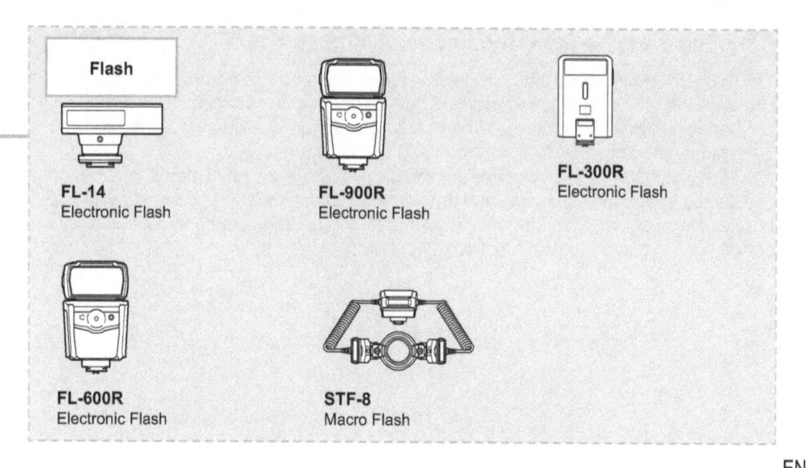
Flash
FL-14
Electronic Flash
FL-900R
Electronic Flash
FL-300R
Electronic Flash
FL-600R
Electronic Flash
STF-8
Macro Flash

Cleaning and storing the camera

Cleaning the camera

Turn off the camera and remove the battery before cleaning the camera.

- Do not use strong solvents such as benzene or alcohol, or a chemically treated cloth.

Exterior:

- Wipe gently with a soft cloth. If the camera is very dirty, soak the cloth in mild soapy water and wring well. Wipe the camera with the damp cloth and then dry it with a dry cloth. If you have used the camera at the beach, use a cloth soaked in clean water and well wrung.

Monitor:

- Wipe gently with a soft cloth.

Lens:

- Blow dust off the lens with a commercially available blower. For the lens, wipe gently with a lens cleaning paper.

Storage

- When not using the camera for a prolonged period, remove the battery and card. Store the camera in a cool, dry place that is well ventilated.
- Insert the battery periodically and test the camera's functions.
- Remove dust and other foreign matter from the body and rear caps before attaching them.
- Attach the body cap to the camera to prevent dust from getting inside when no lens is attached. Be sure to replace the front and rear lens caps before putting the lens away.
- Clean the camera after use.
- Do not store with insect repellent.
- Avoid storing the camera in places where chemicals are treated, in order to protect the camera from corrosion.
- Mold may form on the lens surface if the lens is left dirty.
- Check each part of the camera before use if it has not been used for a long time. Before taking important pictures, be sure to take a test shot and check that the camera works properly.

Cleaning and checking the image pickup device

This camera incorporates a dust reduction function to keep dust from getting on the image pickup device and to remove any dust or dirt from the image pickup device surface with ultrasonic vibrations. The dust reduction function operates when the camera is turned on.
The dust reduction function operates at the same time as the pixel mapping, which checks the image pickup device and image processing circuitry. Since dust reduction is activated every time the camera's power is turned on, the camera should be held upright for the dust reduction function to be effective.

Pixel Mapping - Checking the image processing functions

The pixel mapping feature allows the camera to check and adjust the image pickup device and image processing functions. After using the monitor or taking continuous shots, wait for at least one minute before using the pixel mapping function to ensure that it operates correctly.

1 In Custom Menu J1, select [Pixel Mapping] (P. 122).

2 Press ▷, then press the OK button.

- The [Busy] bar is displayed when pixel mapping is in progress. When pixel mapping is finished, the menu is restored.

- If you accidentally turn the camera off during pixel mapping, start again from Step 1.

After Service

- A warranty is provided at the dealer from whom you purchased the camera. Make sure it includes the name of the store and the purchase date. If one or both of these items are missing, make sure to contact the dealer right away. Carefully read the warranty and store it in a safe place.
- Contact the dealer from whom you purchased the camera or an Olympus service center for after service or if the product malfunctions. If the product malfunctions within one year of the purchase date despite using it according to the instruction manual, Olympus will repair it free of charge based on the warranty.
- Repairs after the warranty period has expired incur fees as a rule.
- After the product has been discontinued, after service will be available for a period of 7 years. However, after service repairs or replacing the product with an equal item (product exchange) at Olympus' discretion is based on the type of malfunction, whether or not replacement parts are available, and the period of time for retaining the parts (parts are generally retained for a period of 7 years after production is discontinued).
- Olympus accepts no responsibility for incidental damages caused by product malfunction (expenses entailed during shooting, and loss of profits from shooting). The customer is responsible for all shipping and handling costs.
- When sending in a product for repairs, make sure it is adequately packed and includes a written notice detailing what needs to be repaired. Make sure to use a package delivery service or registered parcel shipping and obtain a receipt.

8 Information

Shooting tips and information

The camera does not turn on even when a battery is loaded

The battery is not fully charged

- Charge the battery with the charger.

The battery is temporarily unable to function because of the cold

- Battery performance drops at low temperatures. Remove the battery and warm it by putting it in your pocket for a while.

No picture is taken when the shutter button is pressed

The camera has turned off automatically

- If power saving is enabled, the camera will enter sleep mode if no operations are performed for a set period. Press the shutter button halfway to exit the sleep mode.
- The camera automatically enters sleep mode to reduce the drain on the battery if no operations are performed for a set period of time. ☞ [Sleep] (P. 122)
  If no operations are performed for a set time after the camera has entered sleep mode, the camera will turn off automatically. ☞ [Auto Power Off] (P. 122)

The flash is charging

- On the monitor, the ϟ mark blinks when charging is in progress. Wait for the blinking to stop, then press the shutter button.

Unable to focus

- The camera cannot focus on subjects that are too close to the camera or that are not suited to autofocus (the AF confirmation mark will blink in the monitor). Increase the distance to the subject or focus on a high contrast object at the same distance from the camera as your main subject, compose the shot, and shoot.

Subjects that are difficult to focus on

It may be difficult to focus with auto focus in the following situations.

AF confirmation mark is blinking. These subjects are not focused.

Subject with low contrast

Excessively bright light in center of frame

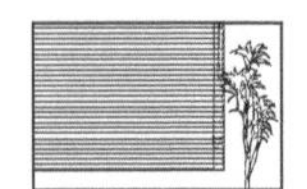

Subject containing no vertical lines

AF confirmation mark lights up but the subject is not focused.

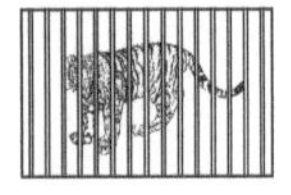

Subjects at different distances

Fast-moving subject

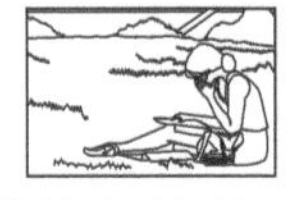

Subject not inside AF area

Noise reduction is activated

- When shooting night scenes, shutter speeds are slower and noise tends to appear in images. The camera activates the noise-reduction process after shooting at slow shutter speeds. During which, shooting is not allowed. You can set [Noise Reduct.] to [Off]. ☞ [Noise Reduct.] (P. 118)

The number of AF targets is reduced

The size and number of the AF target changes depending on the [Digital Tele-converter] (P. 88), [Image Aspect] (P. 54), and group target (P. 39) settings.

The date and time has not been set

The camera is used with the settings at the time of purchase

- The date and time of the camera is not set when purchased. Set the date and time before using the camera. ☞ "Setting the date/time" (P. 19)

The battery has been removed from the camera

- The date and time settings will be returned to the factory default settings if the camera is left without the battery for approximately 1 day. The settings will be canceled more quickly if the battery was only loaded in the camera for a short time before being removed. Before taking important pictures, check that the date and time settings are correct.

Set functions are restored to their factory default settings

When you rotate the mode dial or turn off the power in a shooting mode other than **P**, **A**, **S**, or **M**, functions with changes made to their settings are restored to the factory default settings.

Image taken appears whitish

This may occur when the picture is taken in backlight or semi-backlight conditions. This is due to a phenomenon called flare or ghosting. As far as possible, consider a composition where strong light source is not taken in the picture. Flare may occur even when a light source is not present in the picture. Use a lens hood to shade the lens from the light source. If a lens hood does not have effect, use your hand to shade the lens from the light. ☞ "Interchangeable lenses" (P. 149)

Unknown bright dot(s) appear on the subject in the picture taken

This may be due to stuck pixel(s) on the image pickup device. Perform [Pixel Mapping]. If the problem persists, repeat pixel mapping a few times. ☞ "Pixel Mapping - Checking the image processing functions" (P. 159)

Functions that cannot be selected from menus

Some items may not be selectable from the menus when using the arrow pad.

- Items that cannot be set with the current shooting mode.
- Items that cannot be set because of an item that has already been set:
  Combinations of [♫H] (P. 46, 54) and [Noise Reduct.] (P. 118), etc.

The subject appears distorted

The following functions use an electronic shutter:

movie recording (P. 36), silent mode (P. 47), Pro Capture shooting (P. 48), High Res Shot (P. 48), focus bracketing (P. 94)

This may cause distortion if the subject is moving rapidly or the camera is moved abruptly. Avoid moving the camera abruptly during shooting or use standard sequential shooting.

Lines appear in photographs

The following functions use an electronic shutter, which may result in lines due to flicker and other phenomena associated with fluorescent and LED lighting, an effect that can sometimes be reduced by choosing slower shutter speeds:

movie recording (P. 36), silent mode (P. 47), Pro Capture shooting (P. 48), High Res Shot (P. 48), focus bracketing (P. 94)

Error codes

Monitor indication	Possible cause	Corrective action
No Card	The card is not inserted, or it cannot be recognized.	Insert a card. Or reinsert the card properly.
[1] Card Error	There is a problem with the card in slot 1.	Insert the card again. If the problem persists, format the card. If the card cannot be formatted, it cannot be used.
[2] Card Error	There is a problem with the card in slot 2.	
[1] Write Protect	Writing to the card in slot 1 is prohibited.	The card write-protect switch is set to the "LOCK" side. Release the switch. (P. 147)
[2] Write Protect	Writing to the card in slot 2 is prohibited.	
[1] Card Full	The card is full. No more pictures can be taken or no more information such as print reservation can be recorded.	Replace the card or erase unwanted pictures. Before erasing, download important images to a PC.
	There is no space in the card and print reservation or new images cannot be recorded.	
[2] Card Full	The card is full. No more pictures can be taken or no more information such as print reservation can be recorded.	
	There is no space in the card and print reservation or new images cannot be recorded.	

Monitor indication	Possible cause	Corrective action
Card Setup Clean the contact area of the card with a dry cloth. Clean Card Format Set	Card cannot be read. Card may not have been formatted.	• Select [Clean Card], press the Ⓞ and turn off the camera. Remove the card and wipe the metallic surface with a soft, dry cloth. • Select [Format] ▸ [Yes], and then press the Ⓞ to format the card. Formatting the card erases all data on the card.
1 No Picture	There are no pictures on the card in slot 1.	The card in slot 1 contains no pictures. Record pictures and play back.
2 No Picture	There are no pictures on the card in slot 2.	The card in slot 2 contains no pictures. Record pictures and play back.
1 Picture Error 2 Picture Error	The selected picture cannot be displayed for playback due to a problem with this picture. Or the picture cannot be used for playback on this camera.	Use image processing software to view the picture on a PC. If that cannot be done, the image file is damaged.
1 The Image Cannot Be Edited 2 The Image Cannot Be Edited	Pictures taken with another camera cannot be edited on this camera.	Use image processing software to edit the picture.
1 Cannot Print 2 Cannot Print	Pictures taken with another camera cannot be printed on this camera.	Use image processing software to print the picture.

Monitor indication	Possible cause	Corrective action
°C/°F	The internal temperature of the camera has risen due to sequential shooting.	Turn off the camera and wait for the internal temperature to cool.
Internal camera temperature is too high. Please wait for cooling before camera use.		Wait a moment for the camera to turn off automatically. Allow the internal temperature of the camera to cool before resuming operations.
Battery Empty	The battery is drained.	Charge the battery.
No Connection	The camera is not correctly connected to a computer, printer, HDMI display, or other device.	Reconnect the camera.
No Paper	There is no paper in the printer.	Load some paper in the printer.
No Ink	The printer has run out of ink.	Replace the ink cartridge in the printer.
Jammed	The paper is jammed.	Remove the jammed paper.
Settings Changed	The printer's paper cassette has been removed or the printer has been manipulated while making settings on the camera.	Do not manipulate the printer while making settings on the camera.
Print Error	There is a problem with the printer and/or camera.	Turn off camera and printer. Check the printer and remedy any problems before turning the power on again.
Cannot Print	Pictures recorded on other cameras may not be printed on this camera.	Use a personal computer to print.
The lens is locked. Please extend the lens.	The lens of the retractable lens stays retracted.	Extend the lens.
Please check the status of a lens.	An abnormality has occurred between the camera and the lens.	Turn off the camera, check the connection with the lens, and turn the power on again.

Menu directory

*1: Can be added to [Assign to Custom Mode].
*2: Default can be restored by selecting [Full] for [Reset].
*3: Default can be restored by selecting [Basic] for [Reset].

Shooting Menu

Tab	Function				Default	*1	*2	*3	☞
[Shooting 1]	Reset / Custom Modes				—		✓		86
	Picture Mode				Natural	✓	✓	✓	61, 88
	[Still image quality]				[L]N	✓	✓	✓	55, 88
	Image Aspect				4:3	✓	✓	✓	54
	Digital Tele-converter				Off	✓	✓	✓	88
	[Drive]/[Self-timer]/[Anti-shock]				—	✓	✓	✓	46, 54, 89
		[Drive]/[Self-timer]			[Single]	✓	✓	✓	
		Intrvl. Sh./Time Lapse			Off		✓	✓	90
			Number of Frames		99				
			Start Waiting Time		00:00:01				
			Interval Length		00:00:01				
			Time Lapse Movie		Off				
			Movie Settings	Movie Resolution	FullHD				
				Frame Rate	10fps				
[Shooting 2]	Bracketing				Off	✓	✓	✓	91
		AE BKT			3f 1.0EV				92
		WB BKT	A–B		Off				
			G–M						
		FL BKT			Off				
		ISO BKT			Off				93
		ART BKT			Off				
		Focus BKT			Off	✓	✓	✓	
			Focus Stacking		Off				94
			Set number of shots		99				94
			Set focus differential		5				
			[Flash] Charge Time		0 sec	✓	✓		
	HDR				Off	✓	✓	✓	49, 95
	Multiple Exposure	Number of Frames			Off		✓	✓	95
		Auto Gain			Off				
		Overlay			Off				
	Keystone Comp.				Off	✓	✓	✓	97
	Anti-Shock [♦]/ Silent [♥]	Anti-Shock [♦]			[♦] 0 sec	✓	✓		98
		Silent [♥]			[♥] 0 sec				
		Noise Reduction [♥]			Off				
		Silent [♥] Mode Settings			—	✓	✓		
			[Beep]		Not Allow				
			AF Illuminator		Not Allow				
			Flash Mode		Not Allow				

Tab	Function		Default	*1	*2	*3	☞
2	High Res Shot	High Res Shot	0sec	✔	✔		99
		Charge Time	0 sec				
	RC Mode		Off	✔	✔	✔	99, 153

Video Menu

Tab	Function			Default	*1	*2	*3	☞
(Video)	Mode			P		✔		102
	Specification Settings	(sound icon)		MOV 4K 30p	✔	✔	✔	
		Noise Filter		Normal	✔	✔	✔	
		Picture Mode		Off	✔	✔		
	AF/IS Settings	AF Mode		C-AF	✔	✔	✔	43, 51, 53, 100
		Image Stabilizer		M-IS1	✔	✔	✔	
	Button/Dial/Lever							
		Button Function	Fn1 Function	AF Area Select		✔		100
			Fn2 Function	Multi Function		✔		
			◉ Function	◉ REC		✔		
			AEL/AFL Function	AEL/AFL		✔		
			Function	Peaking		✔		
			Function	Q		✔		
			Function			✔		
			Function	Direct Function		✔		
			▶ Function	Electronic Zoom		✔		
			▼ Function	ISO/ WB		✔		
			B-Fn1 Function	AF Area Select		✔		
			B-Fn2 Function	AEL/AFL		✔		
			PBH Function	Direct Function		✔		
			PBH▶ Function	Electronic Zoom		✔		
			PBH▼ Function	ISO/ WB		✔		
			L-Fn Function	AF Stop		✔		
		Dial Function	P	Exposure / Exposure		✔		
			A	Exposure /FNo.		✔		
			S	Exposure /Shutter		✔		
			M	FNo./ Shutter		✔		

8 Information

Tab	Function			Default	*1	*2	*3	☞
		Fn Lever Function		mode1		✔		100
		Shutter Function				✔		
		Elec. Zoom Speed		Normal		✔		
	Display Settings							
		Control Settings		Live Control, Live SCP		✔		101
		Info Settings		Custom1/ Custom2 (all on except for Movie Effect)		✔		
		Time Code Settings	Time Code Mode	Drop Frame		✔		
			Count Up	Rec Run		✔		
			Starting Time	0:00:00		✔		
		Display Pattern		min	✔	✔		
	Movie			On	✔	✔	✔	103
		Recording Volume	Built-in	±0		✔		
			MIC	±0		✔		
		Volume Limiter		On		✔		
		Wind Noise Reduction		Off		✔		
		Plug-in Power		Off		✔		
		PCM Recorder Link	Camera Rec. Volume	Operative		✔		
			Slate Tone	Off	✔	✔		
			Synchronized ⦿ Rec.	Off	✔	✔		
		Headphone Volume		8	✔	✔		
	HDMI Output		Output Mode	Monitor Mode		✔		101
			REC Bit	Off	✔	✔		
			Time Code	On	✔	✔		

Playback Menu

Tab	Function			Default	*1	*2	*3	☞
		Start		—				80
		BGM		Party Time		✔	✔	
		Slide		All	✔	✔	✔	
		Slide Interval		3 sec	✔	✔		
		Movie Interval		Short	✔	✔		
				On	✔	✔	✔	105
	Edit	Sel. Image	RAW Data Edit	—				105
			JPEG Edit	—				106
			Movie Edit	—				107
				—				83, 107
		Image Overlay		—				107
	Print Order			—				144
	Reset Protect			—				108
	Copy All			—				108
	Connection to Smartphone			—				135

Setup Menu

Tab	Function		Default	*1	*2	*3	☞
	Card Setup		—				110
			—	✔			19
	*		English				109
			±0, ±0, Natural	✔	✔		109
	Rec View		0.5 sec	✔	✔		109
	Wi-Fi Settings	Wi-Fi Connect Settings	Private		✔		137
		Private Password	—				
		Reset share Order	—				138
		Reset Wi-Fi Settings	—				
	Firmware		—				109

* Settings differ depending on the region where the camera is purchased.

✲ Custom Menu

Tab	Function		Default	*1	*2	*3	☞
✲	AF/MF						
A1	AF mode		S-AF	✔	✔	✔	43, 51, 111
	AEL/AFL	S-AF	mode1	✔	✔	✔	111, 123
		C-AF	mode2				
		MF	mode1				
	AF Scanner		mode2	✔	✔	✔	111
	C-AF Lock		±0	✔	✔	✔	
	AF Limiter		Off	✔	✔	✔	
		Distance settings	Setting 1	✔	✔	✔	
		Pri. Rls	On	✔	✔	✔	
	[▦/▪/✣/▦] Settings		All On	✔	✔	✔	
	AF Area Pointer		On1	✔	✔	✔	
A2	AF Targeting Pad		Off	✔	✔	✔	112
	[·:·] Set Home		[▦/▪/✣/▦], ✣		✔	✔	
	[·:·] Custom Settings		Set 1	✔	✔	✔	
		⌃	[▦/▪/✣/▦]	✔	✔	✔	
		⌄	☺	✔	✔	✔	
		⬍	✣	✔	✔	✔	
		◀▶	✣	✔	✔	✔	
	AF Illuminator		On	✔	✔	✔	
	☺Face Priority		☺	✔	✔		40, 112
	AF Focus Adj.		Off	✔	✔	✔	112
A3	Preset MF distance		999.9 m	✔	✔	✔	112
	MF Assist	Magnify	Off	✔	✔		112, 124
		Peaking	Off	✔	✔		
	MF Clutch		Operative	✔	✔	✔	112
	Focus Ring		↻	✔	✔	✔	
	Bulb/Time Focusing		On	✔	✔	✔	
	Reset Lens		On	✔	✔	✔	

Tab	Function			Default	*1	*2	*3	☞
✲	Button/Dial/Lever							
	B	Button Function	Fn1 Function	AF Area Select	✔	✔		66, 113
			Fn2 Function	Multi Function				
			◉ Function	◉ REC				
			AEL/AFL Function	AEL/AFL				
			Function					
			Function					
			Function					
			Function	[·:·]				
			▶ Function	ϟ				
			▼ Function	/				
			B-Fn1 Function	AF Area Select				
			B-Fn2 Function	AEL/AFL				
			PBH Function	[·:·]				
			PBH ▶ Function	ϟ				
			PBH ▼ Function	/				
			L-Fn Function	AF Stop				
		PBH Lock		Off	✔	✔		113
		Dial Function	P	Exposure, Ps	✔	✔		
			A	Exposure, FNo.				
			S	Exposure, Shutter				
			M	FNo., Shutter				
			Menu	◁▷, △▽/Value				
			▶	Q, Prev/Next				
		Dial Direction	Exposure	Dial1	✔	✔		
			Ps	Dial1				
		Fn Lever Settings	Fn Lever Function	mode1		✔		113, 124
			Switch Function	Off		✔		
		Fn Lever/Power Lever		Fn		✔		113
		Elec. Zoom Speed		Normal	✔	✔		

8 Information

Tab	Function			Default	*1	*2	*3	☞
✲	Release/[drive]/Image Stabilizer							
C1	Rls Priority S			Off	✓	✓	✓	114
	Rls Priority C			On	✓	✓	✓	
	[drive]L Settings							
		[drive]/♦[drive]	Max fps	10fps	✓	✓	✓	114
			Frame Count Limiter	Off	✓	✓	✓	
		♥[drive]	Max fps	18fps	✓	✓	✓	
			Frame Count Limiter	Off	✓	✓	✓	
		ProCap	Max fps	18fps	✓	✓	✓	
			Pre-shutter Frames	8 shots	✓	✓	✓	
			Frame Count Limiter	On, 25 shots	✓	✓	✓	
	[drive]H Settings							
		[drive]/♦[drive]	Max fps	15fps	✓	✓	✓	114
			Frame Count Limiter	Off	✓	✓	✓	
		♥[drive]	Max fps	60fps	✓	✓	✓	
			Frame Count Limiter	Off	✓	✓	✓	
		ProCap	Max fps	60fps	✓	✓	✓	
			Pre-shutter Frames	14 shots	✓	✓	✓	
			Frame Count Limiter	On, 25 shots	✓	✓	✓	
C2	[camera] Image Stabilizer			S-IS AUTO	✓	✓	✓	53, 114
	[drive] Image Stabilization			Fps Priority	✓	✓		114
	Half Way Rls With IS			On		✓		
	Lens I.S. Priority			Off	✓	✓	✓	
	Disp/[sound]/PC							
D1	[camera] Control Settings		iAUTO	Live Guide	✓	✓		115, 125
			P/A/S/M	Live SCP	✓	✓		
			ART	Art Menu	✓	✓		
	[index]/Info Settings		[playback] Info	Image Only, Overall	✓	✓	✓	115, 127
			[playback]Q Info	All On		✓		
			LV-Info	Image Only, Custom1 ([histogram]), Custom2 (Level Gauge)	✓	✓		
			[index] Settings	25, My Clips, Calendar	✓	✓		
	Picture Mode Settings			All On	✓	✓		115
	[drive]/⏲ Settings			♥□, [drive]L, [drive]H, ♥[drive]L, ♥[drive]H, ProCapH, ProCapL, [drive], ⏲C, ⏲12s	✓	✓		
	Multi Function Settings			On for settings except [WB/ISO]	✓	✓		

8 Information

Tab		Function		Default	*1	*2	*3	☞
✲	D2	Live View Boost	Manual Shooting	On1	✔	✔	✔	115
			Bulb/Time	On2				
			Live Composite	Off				
			Others	Off				
		Art LV Mode		mode1	✔	✔		
		Frame Rate		Normal	✔	✔	✔	
		LV Close Up Settings	LV Close Up Mode	mode2	✔	✔		116
			Live View Boost	Off	✔	✔		
		⚙ Settings	⚙ Lock	Off	✔	✔		
			Live View Boost	Off	✔	✔		
		Flicker reduction		Auto	✔	✔		
	D3	Grid Settings	Display Color	Preset 1	✔	✔		116
			Displayed Grid	Off	✔	✔		
			Apply Settings to EVF	On	✔	✔		
		Peaking Settings	Peaking Color	Red	✔	✔		
			Highlight Intensity	Normal				
			Image Brightness Adj.	Off				
		Histogram Settings	Highlight	255	✔	✔		
			Shadow	0				
		Mode Guide		Off	✔	✔		
		Selfie Assist		On		✔		116, 129
	D4	■)))		On	✔	✔	✔	117
		HDMI	Output Size	1080p		✔		117, 130
			HDMI Control	Off		✔		
			Output Frame Rate	60p Priority				
		USB Mode		Auto		✔	✔	117
	Exp/ISO/BULB/▣							
	E1	Exposure Shift	▣	±0	✔	✔		117
			◎					
			⊙					
		EV Step		1/3EV	✔	✔	✔	
		ISO Step		1/3EV	✔	✔	✔	
		ISO-Auto Set	Upper Limit / Default	High Limit: 6400 Default: 200	✔	✔	✔	
			Lowest S/S Setting	Auto	✔	✔	✔	
		ISO-Auto		All	✔	✔		
		📷 Noise Filter		Standard	✔	✔	✔	118
		Noise Reduct.		Auto	✔	✔	✔	

8 Information

Tab		Function		Default	*1	*2	*3	☞
✲	E2	Bulb/Time Timer		8min	✓	✓	✓	118
		Bulb/Time Monitor		-7	✓	✓		
		Live Bulb		Off	✓	✓		
		Live Time		0.5 sec	✓	✓		
		Composite Settings		1 sec	✓	✓		30, 118
	E3	Metering		[ESP metering icon]	✓	✓	✓	45, 51, 118
		AEL Metering		Auto	✓	✓	✓	118
		[·:·] Spot Metering	Spot	On	✓	✓	✓	
			Spot Highlight	On	✓	✓	✓	
			Spot Shadow	On	✓	✓	✓	
	ϟ Custom							
	F	ϟX-Sync.		1/250	✓	✓	✓	118, 131
		ϟSlow Limit		1/60	✓	✓	✓	
		[Flash comp icon]+[Exposure comp icon]		Off	✓	✓	✓	39, 60, 118
		ϟ+WB		WB AUTO	✓	✓		119
	[Image quality icon]/WB/Color							
	G	[Image quality icon] Set		[Image quality icon]1 LF, [Image quality icon]2 LN, [Image quality icon]3 MN, [Image quality icon]4 SN	✓	✓	✓	119, 131
		Pixel Count	Middle	3200×2400	✓	✓	✓	
			Small	1280×960				
		Shading Comp.		Off	✓	✓	✓	119
		WB		Auto A±0, G±0	✓	✓	✓	42, 52, 119
		All [WB comp icon]	All Set	—	✓	✓		119
			All Reset	—				
		WB AUTO Keep Warm Color		On	✓	✓	✓	119
		Color Space		sRGB	✓	✓	✓	65, 119
	Record/Erase							
	H1	Card Slot Settings	[Camera icon] Save Settings	Standard	✓	✓		120, 132
			[Camera icon] Save Slot	1	✓	✓		
			[Movie icon] Save Slot	1	✓	✓		
			[Playback icon] Slot	1	✓	✓		
			Assign Save Folder	Do not assign	✓	✓		
		File Name		Reset	✓	✓		120
		Edit Filename		—	✓	✓		
		dpi Settings		350dpi	✓	✓		
		Copyright Settings	Copyright Info.	Off	✓	✓		120
			Artist Name	—				
			Copyright Name	—				
		Lens Info Settings*		Off		✓		120, 132

* [Reset] (Full) and [Reset] (Basic) do not reset the info for individual lenses.

8 Information

Tab	Function			Default	*1	*2	*3	☞
✲	H2	Quick Erase		Off	✓	✓	✓	121
		RAW+JPEG Erase		RAW+JPEG	✓	✓	✓	
		Priority Set		No	✓	✓	✓	
	EVF							
	I	EVF Auto Switch		On		✓		121
		EVF Adjust	EVF Auto Luminance	On	✓	✓		
			EVF Adjust	±0, ☼ ±0				
		EVF Style		Style 3		✓		121, 133
		Info Settings		Basic Information, Custom1 (histogram), Custom2 (level gauge)	✓	✓		121
		EVF Grid Settings	Display Color	Preset 1	✓	✓		
			Displayed Grid	Off	✓	✓		
		Half Way Level		On	✓	✓		
		S-OVF		Off	✓	✓	✓	
	Utility							
	J1	Pixel Mapping		—				122, 159
		Press-and-hold Time	End LV Q	0.7 sec	✓	✓		122
			Reset LV Q Frame	0.7 sec	✓	✓		
			End	0.7 sec	✓	✓		
			Reset Frame	0.7 sec	✓	✓		
			Reset	0.7 sec	✓	✓		
			Reset	0.7 sec	✓	✓		
			Reset	0.7 sec	✓	✓		
			Reset	0.7 sec	✓	✓		
			Reset [·:·]	0.7 sec	✓	✓		
			Call EVF Auto Switch	0.7 sec	✓	✓		
			End ▶Q	0.7 sec	✓	✓		
			Switch /	0.7 sec	✓	✓		
			Insert Slate Tone	0.7 sec	✓	✓		
			End	0.7 sec	✓	✓		
			Reset	0.7 sec	✓	✓		
			Switch Lock	0.7 sec	✓	✓		
			Call BKT Settings	0.7 sec	✓	✓		
		Level Adjust		—		✓		
		Touchscreen Settings		On		✓		
		Menu Recall		Recall	✓	✓		

Tab	Function			Default	*1	*2	*3	☞
⚙	J2	Battery Settings	Battery Priority	PBH Battery	✓	✓		122
			Battery Status	—	✓	✓		
		Backlit LCD		Hold	✓	✓	✓	
		Sleep		1min	✓	✓	✓	
		Auto Power Off		4h	✓	✓	✓	
		Quick Sleep Mode		Off	✓	✓		
			Backlit LCD	8 sec				
			Sleep	10 sec				
		Eye-Fi		On		✓		
		Certification		—				

Default Custom Mode options

Some functions in Custom Modes are preset to different settings from original default settings.

- Settings can be reset to the following settings by selecting [Full] for [Reset] (P. 86) in 📷1 Shooting Menu 1.

Custom Mode C1

Function				Custom Mode C1 options	☞
AF target mode				All targets	39
📷1	▭/◌/▭			▭L	90
	📷◀:			L N+RAW	55
⚙	AF/MF				
	A1	📷 AF mode		C-AF	111
		AF Area Pointer		On2	111
	A2	AF Targeting Pad		On	112
	Disp/ ■))) /PC				
	D2	Live View Boost	Manual Shooting	On1	115
			Bulb/Time	On2	
			Live Composite	Off	
			Others	On1	
		Frame Rate		High	115
	D3	Grid Settings	Display Color	Preset1	116
			Displayed Grid	⊞	

8 Information

Custom Mode C2

Function				Custom Mode C2 options	☞
AF target mode				5-target group	39
AF target position				Center	40
Shooting Menu 1	Sequential/Self-timer/Interval			Pro Cap H (pro capture H)	90
	Image quality			L N+RAW	55
Custom Menu	AF/MF				
	A1	AF mode		S-AF	111
		AF Area Pointer		On1	111
	A2	AF Targeting Pad		On	112
	Disp/ ■))) /PC				
	D2	Live View Boost	Manual Shooting	On1	115
			Bulb/Time	On2	
			Live Composite	Off	
			Others	On1	
		Frame Rate		High	115
	D3	Grid Settings	Display Color	Preset1	116
			Displayed Grid	⊞	

Custom Mode C3

Function				Custom Mode C3 options	☞
AF target mode				[▪] (Single Target)	39
AF target position				Center	40
Shooting Menu 1	Sequential/Self-timer/Interval			Sequential H	90
	Image quality			L N+RAW	55
Custom Menu	AF/MF				
	A1	AF mode		S-AF	111
		AF Area Pointer		On1	111
	A2	AF Targeting Pad		On	112
	Disp/ ■))) /PC				
	D2	Live View Boost	Manual Shooting	On1	115
			Bulb/Time	On2	
			Live Composite	Off	
			Others	On2	
		Frame Rate		Standard	115
	D3	Grid Settings	Display Color	Preset1	116
			Displayed Grid	⊞	
	J2	Quick Sleep Mode		On	122

Specifications

■ Camera

Product type	
Product type	Digital camera with interchangeable Micro Four Thirds Standard lens system
Lens	M.Zuiko Digital, Micro Four Thirds System Lens
Lens mount	Micro Four Thirds mount
Equivalent focal length on a 35mm film camera	Approx. twice the focal length of the lens
Image pickup device	
Product type	4/3" Live MOS sensor
Total no. of pixels	Approx. 21.77 million pixels
No. of effective pixels	Approx. 20.37 million pixels
Screen size	17.3 mm (H) × 13.0 mm (V)
Aspect ratio	1.33 (4:3)
Viewfinder	
Type	Electronic viewfinder with eye sensor
No. of pixels	Approx. 2,360,000 dots
Magnification	100%
Eye point	Approx. 21 mm (–1 m^{-1})
Live view	
Sensor	Uses Live MOS sensor
Magnification	100%
Monitor	
Product type	3.0" TFT color LCD, Vari-angle, touch screen
Total no. of pixels	Approx. 1,040,000 dots (aspect ratio 3: 2)
Shutter	
Product type	Computerized focal-plane shutter
Shutter speed	1/8000 - 60 sec., bulb photography, time photography
Auto focus	
Product type	Hi-Speed Imager AF
Focusing points	121 points
Selection of focusing point	Auto, Optional
Exposure control	
Metering system	TTL metering system (imager metering) Digital ESP metering/Center weighted averaging metering/Spot metering
Metered range	EV –2 - 20 (Equivalent to M.ZUIKO DIGITAL 17mm f2.8, ISO100)
Shooting modes	iAUTO: iAUTO/**P**: Program AE (Program shift can be performed)/ **A**: Aperture priority AE/**S**: Shutter priority AE/**M**: Manual/ **C1**: Custom Mode C1/**C2**: Custom Mode C2/**C3**: Custom Mode C3/ Movie: Movie/**ART**: Art Filter
ISO sensitivity	LOW, 200 - 25600 (1/3, 1 EV step)
Exposure compensation	±5.0EV (1/3, 1/2, 1EV step)
White balance	
Product type	Image pickup device
Mode setting	Auto/Preset WB (7 settings)/Customized WB/One-touch WB (camera can store up to 4 settings)

8 Information

Recording	
Memory	SD, SDHC, SDXC and Eye-Fi UHS-II compatible (slot 1)/UHS-I compatible (slot 2)
Recording system	Digital recording, JPEG (DCF2.0), RAW Data
Applicable standards	Exif 2.3, Digital Print Order Format (DPOF), PictBridge
Sound with still pictures	Wave format
Movie	MPEG-4 AVC/H.264 / Motion JPEG
Audio	Stereo, PCM 48kHz
Playback	
Display format	Single-frame playback/Close-up playback/Index display/Calendar display
Drive	
Drive mode	Single-frame shooting/Sequential shooting/Self-timer
Sequential shooting	Up to 15 fps (□H) Up to 60 fps (♥□H/Pro CapH)
Self-timer	Operation time: 12 sec./2 sec./Customized
Energy saving function	Switch to sleep mode: 1 minute, Power OFF: 4 hours (This function can be customized.)
Flash	
Flash control mode	TTL-AUTO (TTL pre-flash mode)/MANUAL
X-Sync.	1/250 s or slower
Wireless LAN	
Compatible standard	IEEE 802.11b/g/n
External connector	
USB connector (type C)/HDMI micro connector (type D)	
Power supply	
Battery	Lithium-ion Battery ×1
Dimensions/weight	
Dimensions	134.1 mm (W) × 90.9 mm (H) × 68.9 mm (D) (5.3" × 3.6" × 2.7") (excluding protrusions)
Weight	Approx. 574 g (1.3 lb.) (including battery and memory card)
Operating environment	
Temperature	–10 °C - 40 °C (14 °F - 104 °F) (operation)/ –20 °C - 60 °C (–4 °F - 140 °F) (storage)
Humidity	30% - 90% (operation)/10% - 90% (storage)
Splash resistance	Type Equivalent to IEC Standard publication 60529 IPX1 (under OLYMPUS test conditions)

■ Flash

MODEL NO.	FL-LM3
Guide number	9.1 (ISO100•m) 12.7 (ISO200•m)
Firing angle	Covers the picture angle of a 12 mm lens (equivalent to 24 mm in 35 mm format)
Dimensions	Approx. 43.6 mm (W) × 49.4 mm (H) × 39 mm (D) (1.7" × 1.9" × 1.5")
Weight	Approx. 51 g (1.8 oz.)
Splash resistance	Type Equivalent to IEC Standard publication 60529 IPX1 (under OLYMPUS test conditions)

■ Lithium-ion battery

MODEL NO.	BLH-1
Type	Rechargeable Lithium-ion battery
Nominal voltage	DC 7.4 V
Nominal capacity	1720 mAh
No. of charge and discharge times	Approx. 500 times (varies with usage conditions)
Ambient temperature	0 °C - 40 °C (32 °F - 104 °F) (charging)
Dimensions	Approx. 45 mm (W) × 20 mm (H) × 53 mm (D) (1.8" × 0.8" × 2.1")
Weight	Approx. 74 g (2.6 oz.)

■ Lithium-ion charger

MODEL NO.	BCH-1
Rated input	AC 100 V - 240 V (50/60 Hz)
Rated output	DC 8.4 V, 1100 mA
Charging time	Approx. 2 hours (room temperature)
Ambient temperature	0 °C - 40 °C (32 °F - 104 °F) (operation)/ –20 °C - 60 °C (–4 °F - 140 °F) (storage)
Dimensions	Approx. 71 mm (W) × 29 mm (H) × 96 mm (D) (2.8" × 1.1" × 3.8")
Weight (without AC cable)	Approx. 85 g (3.0 oz.)

- The AC cable supplied with this device is for use only with this device and should not be used with other devices. Do not use cables for other devices with this device.

- SPECIFICATIONS ARE SUBJECT TO CHANGE WITHOUT ANY NOTICE OR OBLIGATION ON THE PART OF THE MANUFACTURER.
- Visit our website for the latest specifications.

SAFETY PRECAUTIONS

CAUTION

RISK OF ELECTRIC SHOCK
DO NOT OPEN

CAUTION: TO REDUCE THE RISK OF ELECTRICAL SHOCK, DO NOT REMOVE COVER (OR BACK). NO USER-SERVICEABLE PARTS INSIDE. REFER SERVICING TO QUALIFIED OLYMPUS SERVICE PERSONNEL.

⚠	An exclamation mark enclosed in a triangle alerts you to important operating and maintenance instructions in the documentation provided with the product.
⚠ WARNING	If the product is used without observing the information given under this symbol, serious injury or death may result.
⚠ CAUTION	If the product is used without observing the information given under this symbol, injury may result.
⚠ NOTICE	If the product is used without observing the information given under this symbol, damage to the equipment may result.

WARNING!
TO AVOID THE RISK OF FIRE OR ELECTRICAL SHOCK, NEVER DISASSEMBLE, EXPOSE THIS PRODUCT TO WATER OR OPERATE IN A HIGH HUMIDITY ENVIRONMENT.

General Precautions

Read All Instructions — Before you use the product, read all operating instructions. Save all manuals and documentation for future reference.

Power Source — Connect this product only to the power source described on the product label.

Foreign Objects — To avoid personal injury, never insert a metal object into the product.

Cleaning — Always unplug this product from the wall outlet before cleaning. Use only a damp cloth for cleaning. Never use any type of liquid or aerosol cleaner, or any type of organic solvent to clean this product.

Heat — Never use or store this product near any heat source such as a radiator, heat register, stove, or any type of equipment or appliance that generates heat, including stereo amplifiers.

Attachments — For your safety, and to avoid damaging the product, use only accessories recommended by Olympus.

Location — To avoid damage to the product, mount the product securely on a stable tripod, stand, or bracket.

⚠ WARNING

- **Do not use the camera near flammable or explosive gases.**
- **Rest your eyes periodically when using the viewfinder.**
  Failure to observe this precaution could result in eyestrain, nausea, or sensations similar to motion sickness. The length and frequency of the required rest varies with the individual; use your own judgement. If you feel tired or unwell, avoid using the viewfinder and if necessary consult a physician.
- **Do not use the flash and LED (including AF illuminator) on people (infants, small children, etc.) at close range.**
  - You must be at least 1 m (3 ft.) away from the faces of your subjects. Firing the flash too close to the subject's eyes could cause a momentary loss of vision.
- **Do not look at the sun or strong lights with the camera.**

- **Keep young children, infants away from the camera.**
  - Always use and store the camera out of the reach of young children and infants to prevent the following dangerous situations which could cause serious injury:
    - Becoming entangled in the camera strap, causing strangulation.
    - Accidentally swallowing the battery, cards or other small parts.
    - Accidentally firing the flash into their own eyes or those of another child.
    - Accidentally being injured by the moving parts of the camera.
- **Should you notice that the charger is emitting smoke, heat, or an unusual noise or smell, immediately cease use and unplug the charger from the power outlet, and then contact an authorized distributor or service center.**
- **Stop using the camera immediately if you notice any unusual odors, noise, or smoke around it.**
  - Never remove the batteries with bare hands, which may cause a fire or burn your hands.
- Never hold or operate the camera with wet hands.
  This may cause overheating, exploding, burning, electrical shocks, or malfunctions.
- **Do not leave the camera in places where it may be subject to extremely high temperatures.**
  - Doing so may cause parts to deteriorate and, in some circumstances, cause the camera to catch fire. Do not use the charger if it is covered (such as a blanket). This could cause overheating, resulting in fire.
- **Handle the camera with care to avoid getting a low temperature burn.**
  - When the camera contains metal parts, overheating can result in a low-temperature burn. Pay attention to the following:
    - When used for a long period, the camera will get hot. If you hold on to the camera in this state, a low-temperature burn may be caused.
    - In places subject to extremely cold temperatures, the temperature of the camera's body may be lower than the environmental temperature. If possible, wear gloves when handling the camera in cold temperatures.
- To protect the high-precision technology contained in this product, never leave the camera in the places listed below, no matter if in use or storage:
  - Places where temperatures and/or humidity are high or go through extreme changes. Direct sunlight, beaches, locked cars, or near other heat sources (stove, radiator, etc.) or humidifiers.
  - In sandy or dusty environments.
  - Near flammable items or explosives.
  - In wet places, such as bathrooms or in the rain.
  - In places prone to strong vibrations.
- The camera uses a lithium-ion battery specified by Olympus. Charge the battery with the specified charger. Do not use any other chargers.
- Never incinerate or heat batteries in microwaves, on hot plates, or in pressure vessels, etc.
- Never leave the camera on or near electromagnetic devices.
  This may cause overheating, burning, or exploding.
- Do not connect terminals with any metallic objects.
- Take precautions when carrying or storing batteries to prevent them from coming into contact with any metal objects such as jewelry, pins, fasteners, keys, etc.
  The short circuit may cause overheating, exploding, or burning, which burn or damage you.
- To prevent causing battery leaks or damaging their terminals, carefully follow all instructions regarding the use of batteries. Never attempt to disassemble a battery or modify it in any way, solder, etc.
- If battery fluid gets into your eyes, flush your eyes immediately with clear, cold running water and seek medical attention immediately.
- If you cannot remove the battery from the camera, contact an authorized distributor or service center. Do not try to remove the battery by force.
  Damage to the battery exterior (scratches, etc.) may produce heat or an explosion.
- Always store batteries out of the reach of small children and pets. If they accidentally swallow a battery, seek medical attention immediately.
- To prevent batteries from leaking, overheating, or causing a fire or explosion, use only batteries recommended for use with this product.
- If rechargeable batteries have not been recharged within the specified time, stop charging them and do not use them.
- Do not use batteries with scratches or damage to the casing, and do not scratch the battery.

- Never subject batteries to strong shocks or continuous vibration by dropping or hitting. This may cause exploding, overheating, or burning.
- If a battery leaks, has unusual odor, becomes discolored or deformed, or becomes abnormal in any other way during operation, stop using the camera, and keep away from fire immediately.
- If a battery leaks fluid onto your clothing or skin, remove the clothing and flush the affected area with clean, running cold water immediately. If the fluid burns your skin, seek medical attention immediately.
- The Olympus lithium-ion battery is designed to be used only for the Olympus digital camera. Do not use the battery to other devices.
- **Do not allow children or animals/pets to handle or transport batteries (prevent dangerous behaviour such as licking, putting in mouth or chewing).**

Use Only Dedicated Rechargeable Battery and Battery Charger

We strongly recommend that you use only the genuine Olympus dedicated rechargeable battery and battery charger with this camera.
Using a non-genuine rechargeable battery and/or battery charger may result in fire or personal injury due to leakage, heating, ignition or damage to the battery. Olympus does not assume any liability for accidents or damage that may result from the use of a battery and/or battery charger that are not genuine Olympus accessories.

⚠ CAUTION

- **Do not cover the flash with a hand while firing.**
- Never store batteries where they will be exposed to direct sunlight, or subjected to high temperatures in a hot vehicle, near a heat source, etc.
- Keep batteries dry at all times.
- The battery may become hot during prolonged use. To avoid minor burns, do not remove it immediately after using the camera.
- This camera uses one Olympus lithium-ion battery. Use the specified genuine battery. There is a risk of explosion if the battery is replaced with the incorrect battery type.
- Please recycle batteries to help save our planet's resources. When you throw away dead batteries, be sure to cover their terminals and always observe local laws and regulations.

⚠ NOTICE

- **Do not use or store the camera in dusty or humid places.**
- **Use SD/SDHC/SDXC memory cards or Eye-Fi cards only. Never use other types of cards.**
  If you accidently insert another type of card into the camera, contact an authorized distributor or service center. Do not try to remove the card by force.
- Regularly back up important data to a computer or other storage device to prevent accidental loss.
- OLYMPUS accepts no liability for any loss of data associated with this device.
- Be careful with the strap when you carry the camera. It could easily catch on stray objects and cause serious damage.
- Before transporting the camera, remove a tripod and all other non-OLYMPUS accessories.
- Never drop the camera or subject it to severe shocks or vibrations.
- When attaching the camera to or removing it from a tripod, rotate the tripod screw, not the camera.
- Do not touch electric contacts on cameras.
- Do not leave the camera pointed directly at the sun. This may cause lens or shutter curtain damage, color failure, ghosting on the image pickup device, or may possibly cause fires.
- Do not leave the viewfinder exposed to a strong light source or direct sunlight. The heat may damage the viewfinder.
- Do not push or pull severely on the lens.
- Be sure to remove any water droplets or other moisture from the product before replacing the battery or opening or closing covers.
- Before storing the camera for a long period, remove the batteries. Select a cool, dry location for storage to prevent condensation or mold from forming inside the camera. After storage, test the camera by turning it on and pressing the shutter button to make sure that it is operating normally.
- The camera may malfunction if it is used in a location where it is subject to a magnetic/electromagnetic field, radio waves, or high voltage, such as near a TV set, microwave, video game, loud speakers, large monitor unit, TV/radio tower, or transmission towers. In such cases, turn the camera off and on again before further operation.
- Always observe the operating environment restrictions described in the camera's manual.

- Insert the battery carefully as described in the operating instructions.
- Before loading, always inspect the battery carefully for leaks, discoloration, warping, or any other abnormality.
- Always unload the battery from the camera before storing the camera for a long period.
- When storing the battery for a long period, select a cool location for storage.
- Power consumption by the camera varies depending on which functions are used.
- During the conditions described below, power is consumed continuously and the battery becomes exhausted quickly.
  - The zoom is used repeatedly.
  - The shutter button is pressed halfway repeatedly in shooting mode, activating the auto focus.
  - A picture is displayed on the monitor for an extended period of time.
  - The camera is connected to a printer.
- Using an exhausted battery may cause the camera to turn off without displaying the battery level warning.
- If the battery's terminals get wet or greasy, camera contact failure may result. Wipe the battery well with a dry cloth before use.
- Always charge a battery when using it for the first time, or if it has not been used for a long period.
- When operating the camera with battery power at low temperatures, try to keep the camera and spare battery as warm as possible. A battery that has run down at low temperatures may be restored after it is warmed at room temperature.
- Before going on a long trip, and especially before traveling abroad, purchase extra batteries. A recommended battery may be difficult to obtain while traveling.

Using the wireless LAN function

- **Turn off the camera in hospitals and other locations where medical equipment is present.**
  The radio waves from the camera may adversely affect medical equipment, causing a malfunction that results in an accident.
- **Turn off the camera when onboard aircraft.**
  Using wireless devices onboard may hinder safe operation of the aircraft.

Monitor

- Do not push the monitor forcibly; otherwise the image may become vague, resulting in a playback mode failure or damage to the monitor.
- A strip of light may appear on the top/bottom of the monitor, but this is not a malfunction.
- When a subject is viewed diagonally in the camera, the edges may appear zigzagged on the monitor. This is not a malfunction; it will be less noticeable in playback mode.
- In places subject to low temperatures, the monitor may take a long time to turn on or its color may change temporarily.
  When using the camera in extremely cold places, it is a good idea to occasionally place it in a warm place. The monitor exhibiting poor performance due to low temperatures will recover in normal temperatures.
- The monitor of this product is manufactured with high-quality accuracy, however, there may be a stuck or dead pixel on the monitor. These pixels do not have any influence on the image to be saved. Because of the characteristics, the unevenness of the color or brightness may also be found depending on the angle, but this is due to the structure of the monitor. This is not a malfunction.

Legal and Other Notices

- Olympus makes no representations or warranties regarding any damages, or benefit expected by using this unit lawfully, or any request from a third person, which are caused by the inappropriate use of this product.
- Olympus makes no representations or warranties regarding any damages or any benefit expected by using this unit lawfully which are caused by erasing picture data.

Disclaimer of Warranty

- Olympus makes no representations or warranties, either expressed or implied, by or concerning any content of these written materials or software, and in no event shall be liable for any implied warranty of merchantability or fitness for any particular purpose or for any consequential, incidental or indirect damages (including but not limited to damages for loss of business profits, business interruption and loss of business information) arising from the use or inability to use these written materials or software or equipment. Some countries do not allow the exclusion or limitation of liability for consequential or incidental damages or of the implied warranty, so the above limitations may not apply to you.
- Olympus reserves all rights to this manual.

Warning

Unauthorized photographing or use of copyrighted material may violate applicable copyright laws. Olympus assumes no responsibility for unauthorized photographing, use or other acts that infringe upon the rights of copyright owners.

Copyright Notice

FCC Notice

This equipment has been tested and found to comply with the limits for a Class B digital device, pursuant to part 15 of the FCC Rules. These limits are designed to provide reasonable protection against harmful interference in a residential installation. This equipment generates, uses and can radiate radio frequency energy and, if not installed and used in accordance with the instructions, may cause harmful interference to radio communications. However, there is no guarantee that interference will not occur in a particular installation. If this equipment does cause harmful interference to radio or television reception, which can be determined by turning the equipment off and on, the user is encouraged to try to correct the interference by one or more of the following measures:

- Reorient or relocate the receiving antenna.
- Increase the separation between the equipment and receiver.
- Connect the equipment into an outlet on a circuit different from that to which the receiver is connected.
- Consult the dealer or an experienced radio/TV technician for help.
- Only the OLYMPUS-supplied USB cable should be used to connect the camera to USB enabled personal computers.

FCC/IC Caution

Changes or modifications not expressly approved by the party responsible for compliance could void the user's authority to operate the equipment.

This transmitter must not be co-located or operated in conjunction with any other antenna or transmitter.

This equipment complies with FCC radiation exposure limits set forth for an uncontrolled environment and meets the FCC radio frequency (RF) Exposure Guidelines. This equipment has very low levels of RF energy that are deemed to comply without testing of specific absorption rate (SAR).

The available scientific evidence does not show that any health problems are associated with using low power wireless devices. There is no proof, however, that these low power wireless devices are absolutely safe. Low power Wireless devices emit low levels of radio frequency energy (RF) in the microwave range while being used. Whereas high levels of RF can produce health effects (by heating tissue), exposure of low-level RF that does not produce heating effects causes no known adverse health effects. Many studies of low-level RF exposures have not found any biological effects. Some studies have suggested that some biological effects might occur, but such findings have not been confirmed by additional research. IM002 has been tested and found to comply with IC radiation exposure limits set forth for an uncontrolled environment and meets RSS-102 of the IC radio frequency (RF) Exposure rules.

For customers in North America, Central America, South America and the Caribbean

Declaration of Conformity
Model Number : IM002
Trade Name : OLYMPUS
Responsible Party : **OLYMPUS AMERICA INC.**
Address : 3500 Corporate Parkway, P. O. Box 610, Center Valley, PA 18034-0610, USA
Telephone Number : 484-896-5000
Tested To Comply With FCC Standards
FOR HOME OR OFFICE USE

This device complies with Part 15 of FCC Rules and Industry Canada licence-exempt RSS standard(s). Operation is subject to the following two conditions:

(1) This device may not cause harmful interference.
(2) This device must accept any interference received, including interference that may cause undesired operation.

CAN ICES-3(B)/NMB-3(B)

OLYMPUS AMERICAS LIMITED WARRANTY - OLYMPUS AMERICA INC. PRODUCTS

Olympus warrants that the enclosed Olympus® imaging product(s) and related Olympus® accessories (individually a "Product" and collectively the "Products") will be free from defects in materials and workmanship under normal use and service for a period of one (1) year from the date of purchase.

If any Product proves to be defective within the one-year warranty period, the customer must return the defective Product to the authorized Olympus Repair Service Center designated by Olympus, following the procedure set forth below (See "WHAT TO DO WHEN SERVICE IS NEEDED").

Olympus, at its sole discretion, will repair, replace, or adjust the defective Product at Olympus's cost, provided that an Olympus investigation and factory inspection disclose that (a) such defect developed under normal and proper use and (b) the Product is covered under this limited warranty.

Repair, replacement, or adjustment of defective Products shall be Olympus's sole obligation and the customer's sole remedy hereunder. Repair or replacement of a Product shall not extend the warranty period provided herein, unless required by law.

Except where prohibited by law, the customer is liable and shall pay for shipment of the Products to the designated Olympus Repair Service Center. Olympus shall not be obligated to perform preventive maintenance, installation, deinstallation, or maintenance.

Olympus reserves the right to (i) use reconditioned, refurbished, and/or serviceable used parts (that meet Olympus's quality assurance standards) for warranty or any other repairs and (ii) make any internal or external design and/or feature changes on or to its products without any liability to incorporate such changes on or to the Products.

WHAT IS NOT COVERED BY THIS LIMITED WARRANTY

Excluded from this limited warranty and not warranted by Olympus in any fashion, either express, implied, or by statute, are:

(a) products and accessories not manufactured by Olympus and/or not bearing the "OLYMPUS" brand label (the warranty coverage for products and accessories of other manufacturers, which may be distributed by Olympus, is the responsibility of the respective manufacturer of such products and accessories in accordance with the terms and duration of such manufacturers' warranties);
(b) any Product which has been disassembled, repaired, tampered with, altered, changed, or modified by persons other than Olympus's own authorized service personnel unless repair by others is made with the written consent of Olympus;
(c) defects or damage to the Products resulting from wear, tear, misuse, abuse, negligence, sand, liquids, impact, improper storage, nonperformance of scheduled operator and maintenance items, battery leakage, use of non-"OLYMPUS" brand accessories, consumables, or supplies, or use of the Products in combination with non-compatible devices;
(d) software programs;
(e) supplies and consumables (including but not limited to lamps, ink, paper, film, prints, negatives, cables and batteries); and/or

(f) Products which do not contain a validly placed and recorded Olympus serial number, unless they are a model on which Olympus does not place and record serial numbers.
(g) Products shipped, delivered, purchased, or sold from dealers located outside of North America, Central America, South America and the Caribbean; and/or
(h) Products that are not intended or authorized to be sold in North America, South America, Central America, or the Caribbean (ie. Gray Market Goods).

WARRANTY DISCLAIMER; LIMITATION OF DAMAGES; AFFIRMATION OF ENTIRE WARRANTY AGREEMENT; INTENDED BENEFICIALLY

EXCEPT FOR THE LIMITED WARRANTY SET FORTH ABOVE, OLYMPUS MAKES NO AND DISCLAIMS ALL OTHER REPRESENTATIONS, GUARANTIES, CONDITIONS, AND WARRANTIES CONCERNING THE PRODUCTS, WHETHER DIRECT OR INDIRECT, EXPRESS OR IMPLIED, OR ARISING UNDER ANY STATUTE, ORDINANCE, COMMERCIAL USAGE OR OTHERWISE, INCLUDING BUT NOT LIMITED TO ANY WARRANTY OR REPRESENTATION AS TO THE SUITABILITY, DURABILITY, DESIGN, OPERATION, OR CONDITION OF THE PRODUCTS (OR ANY PART THEREOF) OR THE MERCHANTABILITY OF THE PRODUCTS OR THEIR FITNESS FOR A PARTICULAR PURPOSE, OR RELATING TO THE INFRINGEMENT OF ANY PATENT, COPYRIGHT, OR OTHER PROPRIETARY RIGHT USED OR INCLUDED THEREIN.

IF ANY IMPLIED WARRANTIES APPLY AS A MATTER OF LAW, THEY ARE LIMITED IN DURATION TO THE LENGTH OF THIS LIMITED WARRANTY.

SOME STATES MAY NOT RECOGNIZE A DISCLAIMER OR LIMITATION OF WARRANTIES AND/OR LIMITATION OF LIABILITY SO THE ABOVE DISCLAIMERS AND EXCLUSIONS MAY NOT APPLY.

THE CUSTOMER MAY ALSO HAVE DIFFERENT AND/OR ADDITIONAL RIGHTS AND REMEDIES THAT VARY FROM STATE TO STATE.

THE CUSTOMER ACKNOWLEDGES AND AGREES THAT OLYMPUS SHALL NOT BE RESPONSIBLE FOR ANY DAMAGES THAT THE CUSTOMER MAY INCUR FROM DELAYED SHIPMENT, PRODUCT FAILURE, PRODUCT DESIGN, SELECTION, OR PRODUCTION, IMAGE OR DATA LOSS OR IMPAIRMENT OR FROM ANY OTHER CAUSE, WHETHER LIABILITY IS ASSERTED IN CONTRACT, TORT (INCLUDING NEGLIGENCE AND STRICT PRODUCT LIABILITY) OR OTHERWISE. IN NO EVENT SHALL OLYMPUS BE LIABLE FOR ANY INDIRECT, INCIDENTAL, CONSEQUENTIAL OR SPECIAL DAMAGES OF ANY KIND (INCLUDING WITHOUT LIMITATION LOSS OF PROFITS OR LOSS OF USE), WHETHER OR NOT OLYMPUS SHALL BE OR SHOULD BE AWARE OF THE POSSIBILITY OF SUCH POTENTIAL LOSS OR DAMAGE.

Representations and warranties made by any person, including but not limited to dealers, representatives, salespersons, or agents of Olympus, which are inconsistent or in conflict with or in addition to the terms of this limited warranty, shall not be binding upon Olympus unless reduced to writing and approved by an expressly authorized officer of Olympus.

This limited warranty is the complete and exclusive statement of warranty which Olympus agrees to provide with respect to the Products and it shall supersede all prior and contemporaneous oral or written agreements, understandings, proposals, and communications pertaining to the subject matter hereof.

This limited warranty is exclusively for the benefit of the original customer and cannot be transferred or assigned.

WHAT TO DO WHEN SERVICE IS NEEDED

The customer must contact the designated Olympus Consumer Support Team for your region to coordinate the submission of your Product for repair service. To contact your Olympus Consumer Support Team in your region please visit or call the following:

Canada:

www.olympuscanada.com/repair /
1-800-622-6372

United States:

www.olympusamerica.com/repair /
1-800-622-6372

Latin America:

www.olympusamericalatina.com

The customer must copy or transfer any image or other data saved on a Product to another image or data storage medium prior to sending the Product to Olympus for for repair service.

IN NO EVENT SHALL OLYMPUS BE RESPONSIBLE FOR SAVING, KEEPING OR MAINTAINING ANY IMAGE OR DATA SAVED ON A PRODUCT RECEIVED BY IT FOR SERVICE, OR ON ANY FILM CONTAINED WITHIN A PRODUCT RECEIVED BY IT FOR SERVICE, NOR SHALL OLYMPUS BE RESPONSIBLE FOR ANY DAMAGES IN THE EVENT ANY IMAGE OR DATA IS LOST OR IMPAIRED WHILE SERVICE IS BEING PERFORMED (INCLUDING, WITHOUT

LIMITATION, DIRECT, INDIRECT, INCIDENTAL, CONSEQUENTIAL OR SPECIAL DAMAGES, LOSS OF PROFITS OR LOSS OF USE), WHETHER OR NOT OLYMPUS SHALL BE OR SHOULD BE AWARE OF THE POSSIBILITY OF SUCH POTENTIAL LOSS OR IMPAIRMENT.

The customer should package the Product carefully using ample padding material to prevent damage in transit. Once the Product is properly packaged, ship the package to Olympus or the Olympus Authorized Repair Service Center location as instructed by the respective Olympus Consumer Support Team.

When sending Products for repair service, your package should include the following:

1) Sales receipt showing date and place of purchase. Handwritten receipts will not be accepted;
2) Copy of this limited warranty **bearing the Product serial number corresponding to the serial number on the Product** (unless it is a model on which Olympus does not place and record serial numbers);
3) A detailed description of the problem; and
4) Sample prints, negatives, digital prints (or files on disk) if available and related to the problem.

KEEP COPIES OF ALL DOCUMENTS. Neither Olympus nor an Olympus Authorized Repair Service Center will be responsible for documents that are lost or destroyed in transit.

When service is completed, the Product will be returned to you postage prepaid.

PRIVACY

Any information provided by you to process your warranty claim shall be kept confidential and will only be used and disclosed for the purposes of processing and performing warranty repair services.

For customers in Europe

Simple Declaration of Comformity

Hereby, OLYMPUS CORPORATION declares that the radio equipment type IM002 is in compliance with Directive 2014/53/EU.

The full text of the EU declaration of conformity is available at the following internet address: http://www.olympus-europa.com/

This symbol [crossed-out wheeled bin WEEE Annex IV] indicates separate collection of waste electrical and electronic equipment in the EU countries.

Please do not throw the equipment into the domestic refuse.

Please use the return and collection systems available in your country for the disposal of this product.

This symbol [crossed-out wheeled bin Directive 2006/66/EC Annex II] indicates separate collection of waste batteries in the EU countries.

Please do not throw the batteries into the domestic refuse.

Please use the return and collection systems available in your country for the disposal of the waste batteries.

Provisions of warranty

In the unlikely event that your product proves to be defective, although it has been used properly (in accordance with the written Instruction Manual supplied with it), during the applicable national warranty period and has been purchased from an authorized Olympus distributor within the business area of OLYMPUS EUROPA SE & Co. KG as stipulated on the website: http://www.olympus-europa.com, it will be repaired, or at Olympus's option replaced, free of charge. In order to enable Olympus to provide you with the requested warranty services to your full satisfaction and as fast as possible, please note the information and instructions listed below:

1. To claim under this warranty please follow the instructions on http://consumer-service.olympus-europa.com for registration and tracking (this service is not available in all countries) or take the product, the corresponding original invoice or purchase receipt and the completed Warranty Certificate to the dealer where it was purchased or any other Olympus service station within the business area of OLYMPUS EUROPA SE & Co. KG as stipulated on the website: http://www.olympus-europa.com, before the end of the applicable national warranty period.
2. Please make sure your Warranty Certificate is duly completed by Olympus or an authorized dealer or Service center. Therefore, please make sure that your name, the name of the dealer, the serial number and the year, month and date of purchase are all completed or the original invoice or the sales receipt (indicating the dealer's name, the date of purchase and product type) is attached to your Warranty Certificate.
3. Since this Warranty Certificate will not be re-issued, keep it in a safe place.
4. Please note that Olympus will not assume any risks or bear any costs incurred in transporting the product to the dealer or Olympus authorized service station.
5. This Warranty does not cover the following and you will be required to pay a repair charge, even for defects occurring within the warranty period referred to above.
   a. Any defect that occurs due to mishandling (such as an operation performed that is not mentioned in the Instruction Manual, etc.)
   b. Any defect that occurs due to repair, modification, cleaning, etc. performed by anyone other than Olympus or an Olympus authorized service station.
   c. Any defect or damage that occurs due to transport, a fall, shock, etc. after purchase of the product.
   d. Any defect or damage that occurs due to fire, earthquake, flood damage, thunderbolt, other natural disasters, environmental pollution and irregular voltage sources.
   e. Any defect that occurs due to careless or improper storage (such as keeping the product under conditions of high temperature and humidity, near insect repellents such as naphthalene or harmful drugs, etc.), improper maintenance, etc.
   f. Any defect that occurs due to exhausted batteries, etc.
   g. Any defect that occurs due to sand, mud, water etc. entering the inside of the product casing.
6. Olympus's sole liability under this Warranty shall be limited to repairing or replacing the product. Any liability under the Warranty for indirect or consequential loss or damage of any kind incurred or suffered by the customer due to a defect of the product, and in particular any loss or damage caused to any lenses, films, other equipment or accessories used with the product or for any loss resulting from a delay in repair or loss of data, is excluded. Compelling regulations by law remain unaffected by this.

For customers in Thailand

This telecommunication equipment is in compliance with NTC requirements.

For customers in Mexico

The operation of this equipment is subject to the following two conditions:
(1) it is possible that this equipment or device may not cause harmful interference, and (2) this equipment or device must accept any interference, including interference that may cause undesired operation.

For customer in Singapore

Complies with
IDA Standards
DB104634

Trademarks

- Microsoft and Windows are registered trademarks of Microsoft Corporation.
- Macintosh is a trademark of Apple Inc.
- SDXC Logo is a trademark of SD-3C, LLC.
- Eye-Fi is a trademark of Eye-Fi, Inc.
- "Shadow Adjustment Technology" function contains patented technologies from Apical Limited.

- Micro Four Thirds, Four Thirds, and the Micro Four Thirds and Four Thirds logos are trademarks or registered trademarks of the OLYMPUS CORPORATION in Japan, the United States, the countries of the European Union, and other countries.
- Wi-Fi is a registered trademark of the Wi-Fi Alliance.
- The Wi-Fi CERTIFIED logo is a certification mark of the Wi-Fi Alliance.

- The standards for camera file systems referred to in this manual are the "Design Rule for Camera File System/DCF" standards stipulated by the Japan Electronics and Information Technology Industries Association (JEITA).
- All other company and product names are registered trademarks and/or trademarks of their respective owners.

The software in this camera may include third party software. Any third party software is subject to the terms and conditions, imposed by the owners or licensors of that software, under which software is provided to you.

Those terms and other third party software notices, if any, may be found in the software notice PDF file stored at

http://www.olympus.co.jp/en/support/imsg/digicamera/download/notice/notice.cfm

10 Additions/modifications by firmware update

The following functions are added/modified by firmware update.

Additions/modifications by the firmware version 2.0

Information displays while shooting

Monitor display during still photography

The battery level icon has been modified.
Icon displays for Pro Capture shooting, Flicker Scan, and Fisheye Compensation shooting have been added.

① Battery level display
② Pro Capture shooting.....................P.191
③ Fisheye Compensation shooting ...P.193
④ Flicker Scan..................................P.193

Pro Capture shooting (buffering capacity increased)

[Pre-shutter Frames] in Pro Capture shooting (P.48) has been extended to up to 35 frames.
[Pre-shutter Frames] can be set in [Pro Cap] of [L Settings] or [H Settings] (P. 114) in Custom Menu.

Pro Capture shooting (supported lenses added)

Lenses that support Pro Capture shooting (P.48) have been added.
See the OLYMPUS website for information on the lenses that can be used with Pro Capture shooting.

Types of art filters ([Bleach Bypass] added)

[Bleach Bypass] is added to the art filters (P. 33).

Bleach Bypass I/II	The "bleach bypass" effect, which you may recognize from motion pictures and the like, can be used to great effect in shots of streetscapes or metal objects.

"II" is an alternate version of the original (I).

AF target mode ([Small Target] added)

[Small Target] is added to the AF target mode (P. 39).

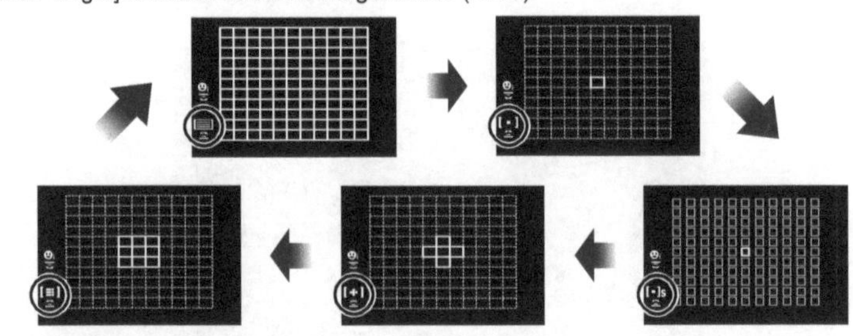

[▦] (All Targets)	The camera automatically chooses from the full set of focus targets.
[▪] (Single Target)	You can choose a single AF target.
[▪]s (Small Target)	The AF target can be reduced in size.
[·:·] (5-Target Group)	The camera automatically chooses from the targets in the selected five-target group.
[⁝⁝⁝] (9-Target Group)	The camera automatically chooses from the targets in the selected nine-target group.

▶ Q Default Setting

A zoom ratio setting for close-up image playback (P.79) has been added.
The following option is added in the custom menus (**MENU ➔ ✲ ➔ D2**) (P.115).

Option	Description
▶ Q **Default Setting**	Choose the zoom ratio from [Recently], [Equally Value], [×2], [×3], [×5], [×7], [×10] or [×14]. When [Equally Value] is selected, 1:1 is displayed on the monitor.

Flicker Scan

This function lets you shoot images with less flicker from indoor LED lamps, by changing the shutter speed incrementally while you check the degree of flicker on the live-view display.
While Flicker Scan is activated, if you want to use shooting functions such as aperture adjustment or exposure compensation, press the **INFO** button to switch to another setting screen.
To return to the Flicker Scan screen, press the **INFO** button repeatedly until the screen switches.

- Using Zoom AF display (P.41) makes the degree of flicker easier to assess.
- The range of shutter speeds that can be set is smaller when Flicker Scan is activated.

During still image shooting

The following option is added in the custom menus (**MENU → ✲ → E2**) (P.118).
Flicker Scan is available during Silent [♥] shooting, Pro Capture shooting, or High Res Shot when the mode dial is set to **S** or **M**.

Option	Description
Flicker Scan	Set the option to [On], and use the front dial/rear dial or △▽ on the arrow pad to adjust the shutter speed until flicker is reduced.

During movie recording

The following option is added in Video Menu (**MENU → → Mode Settings**) (P.100).
Flicker Scan is available when the mode dial is set to , and Mode is set to **S** or **M**.

Option	Description
Flicker Scan	Set the option to [On], and use the front dial/rear dial or △▽ on the arrow pad to adjust the shutter speed until flicker is reduced.

Fisheye Compensation

This function compensates for the distortion from a fisheye lens, allowing you to shoot images resembling those taken with a super-wide-angle lens.
This function can only be set when you attach a supported fisheye lens*1.
The following option is added in the custom menus (**MENU → ✲ → J1**) (P.122).

Option	Description
Fisheye Compensation	Set the option to [On], and press ▷ on the arrow pad to set the detailed options. The angle (1 to 3), and [On]/[Off] for Correction can be set.

*1 M.ZUIKO DIGITAL ED 8mm f1.8 Fisheye PRO supports this function (as of the end of February 2018).

Additions/modifications by firmware update

10

Button Function

During still image shooting

The following options are added in the custom menus (**MENU ➔ ✲ ➔ B ➔ Button Function**) (P.66).

Option	Description
Flicker Scan	Press the button to activate Flicker Scan. To deactivate the function, press and hold down the button. Use the front dial (front dial)/rear dial (rear dial) or △▽ on the arrow pad to adjust the shutter speed. When [On] is selected, press the button again to switch the information display. • Flicker Scan is available during Silent [♥] shooting, Pro Capture shooting, or High Res Shot when the mode dial is set to **S** or **M**.
Fisheye Compensation	Press the button to activate Fisheye Compensation. Press the button again to cancel Fisheye Compensation. Use the front dial (front dial) or rear dial (rear dial) while holding down the button to select the compensation level.

During movie recording

The following option is added in Video Menu (**MENU ➔ 🎥 ➔ 🎥 Button/Dial/Lever ➔ 🎥 Button Function**) (P.100).

Option	Description
Flicker Scan	Press the button to activate Flicker Scan. To deactivate the function, press and hold down the button. Use the front dial (front dial)/rear dial (rear dial) or △▽ on the arrow pad to adjust the shutter speed. When [On] is selected, press the button again to switch the information display. • Flicker Scan is available when the mode dial is set to 🎥, and 🎥 Mode is set to **S** or **M**.

Default settings

The default settings for new functions and the modified default settings are as follows.

*1: Can be added to [Assign to Custom Mode].

*2: Default can be restored by selecting [Full] for [Reset].

*3: Default can be restored by selecting [Basic] for [Reset].

🎥 Video Menu

Tab	Function		Default	*1	*2	*3	☞
🎥	🎥 Mode Settings	🎥 Mode	P		✓		102
		🎥 Flicker Scan	Off	✓	✓	✓	193

🎥 Custom Menu

Tab	Function		Default	*1	*2	*3	☞
✲	D2	▶ 🔍 Default Setting	Recently	✓	✓		192
	E2	📷 Flicker Scan	Off	✓	✓	✓	193

Index

Symbols

A

B

Q

R

S

T

U

V

W

Z

OLYMPUS KOREA CO., LTD.
Olympus Tower A, 446, Bongeunsa-ro, Gangnam-gu, Seoul, Korea, 06153
Tel. 1544-3200
Email: hotline.okr@olympus-ap.com
http://www.olympus.co.kr

OLYMPUS (MALAYSIA) Sdn Bhd
512, 5th Floor, Block D, Kelana Square 17, Jalan SS 7/26, Kelana Jaya,
47301 Petaling Jaya, Selangor, Malaysia
Tel: (603) 7806 2173
Fax: (603) 7803 7164
E-mail: service.oml@olympus-ap.com
http://www.olympus.com.my

OLYMPUS (Thailand) CO., LTD.
23/112 Sorachai Building, 27th Floor, Soi Sukhumvit 63 (Ekamai)
Sukhumvit road Klongton Nua, Wattana, Bangkok 10110 Thailand
Tel: 662 787 8200
E-mail：imaging.oth@olympus-ap.com

date of issue 2016.9

http://www.olympus.com/

OLYMPUS AMERICA INC.

3500 Corporate Parkway, P.O. Box 610, Center Valley, PA 18034-0610, U.S.A. Tel. 484-896-5000

Technical Support (U.S.A. / Canada)
24/7 online automated help:
http://www.olympusamerica.com/support
Phone customer support:
Tel. 1-800-260-1625 (Toll-free)

Our phone customer support is available from 9 am to 9 pm (Monday to Friday) ET
http://olympusamerica.com/contactus
Olympus software updates can be obtained at:
http://www.olympusamerica.com/digital

OLYMPUS EUROPA SE & CO. KG

Premises: Consumer Product Division
Wendenstrasse 14-18, 20097
Hamburg, Germany
Tel: +49 40-23 77 3-0 / Fax: +49 40-23 07 61

Goods delivery: Modul H, Willi-Bleicher Str. 36,
52353 Düren, Germany
Mailing address: Postfach 10 49 08,
20034 Hamburg, Germany

European Technical Customer Support:

Please visit our homepage **http://www.olympus-europa.com** or call our TOLL FREE NUMBER* :
00800 - 67 10 83 00
for Austria, Belgium, Denmark, Finland, France, Germany, Luxemburg, Netherlands, Norway, Poland, Portugal, Spain, Sweden, Switzerland, United Kingdom.
810-800 67 10 83 00 for Russia
800 167 777 for Czech Republic
* Please note some phone service providers do not permit access to 00800 numbers or require an additional prefix. Charges may apply here. Please contact your service provider directly for more details.
For all not listed European Countries and in case that you can't get connected to the above mentioned number, please make use of the following CHARGED NUMBERS **+49 40 - 237 73 899**

OLYMPUS (CHINA) CO., LTD.

Customer Support: 400-650-0303 Homepage: http://olympus-imaging.cn

Customer Service Center:
10F, K, Wah Centre, 1010 Huaihai Road(M), Xuhui District, Shanghai
Zip: 200031

OLYMPUS HONG KONG AND CHINA LIMITED

Digital Camera Repair Centre
L4207, Office Tower, Langham Place, 8 Argyle Street, Mongkok, Kowloon, Hong Kong
Customer Hotline: +852-2376-2150 Fax: +852-2375-0630
E-mail: cs.ohc@olympus-ap.com
http://www.olympus.com.hk

WD169003